Rainer Flindt · Amazing Numbers in Biology

D1211825

QH323.5 .F58 2006

Flindt, Rainer.

Amazing numbers in
 biology
      c2006.

2006 10 19

170101

Rainer Flindt

Amazing Numbers in Biology

Translated by Neil Solomon

PROPERTY OF
SENECA COLLEGE
LIBRARIES
@ YORK CAMPUS

WITHDRAWN

 Springer

Professor Dr. Rainer Flindt
Dietrich-Bonhoeffer-Weg 16
71462 Ludwigsburg
Germany

Translated by:

Neil Solomon
Dantestr. 53
69115 Heidelberg
Germany

This is the translation of the 6th German Edition of

"Biologie in Zahlen"
© 2003 Elsevier GmbH, Spektrum Akademischer Verlag, Heidelberg
(ISBN 3-8274-1171-8)

The translation was undertaken by Springer Verlag

Library of Congress Control Number: 2005938321

ISBN-10 3-540-30146-1 Springer Berlin Heidelberg New York
ISBN-13 978-3-540-30146-2

This work is subject to copyright. All rights reserved, whether the whole or part of the material is concerned, specifically the rights of translation, reprinting, reuse of illustrations, recitation, broadcasting, reproduction on microfilm or in any other way, and storage in data banks. Duplication of this publication or parts thereof is permitted only under the provisions of the German Copyright Law of September 9, 1965, in its current version, and permission for use must always be obtained from Springer-Verlag. Violations are liable for prosecution under the German Copyright Law.

Springer-Verlag is a part of Springer Science + Business Media
springeronline.com

© Springer-Verlag Berlin Heidelberg 2006
Printed in Germany

The use of general descriptive names, registered names, trademarks, etc. in this publication does not imply, even in the absence of a specific statement, that such names are exempt from the relevant protective laws and regulations and therefore free for general use.

Editor: Dr. Dieter Czeschlik, Heidelberg, Germany
Desk Editor: Dr. Jutta Lindenborn, Heidelberg, Germany
Cover design: Design & Production, Heidelberg
Typesetting and Production: Friedmut Kröner, Heidelberg
Printed on acid-free paper – 39/3152-YK – 5 4 3 2 1 0

Contents

Introduction to the 6th German Edition 1

Part 1: Zoology

1.1   **General: Size, Weight, Age ...** 4
1.1.0   Numbers that Amaze 4
1.1.1   Approximate Numbers of Species in Selected Major Animal Taxa 5
1.1.2   Maximum Age of Selected Animals 6
1.1.3   Largest Representatives of Selected Animal Taxa
       (Excluding Vertebrates and Insects) 8
1.1.4   The Largest Insects 10
1.1.5   Size of Populations in Colony-Building Insects 11
1.1.6   Size and Weight of Selected Fish 12
1.1.7   Maximum Length of Selected Amphibians 13
1.1.8   Maximum Length and Weight of Selected Reptiles 14
1.1.9   Weight and Wingspan of Selected Birds 15
1.1.10 Maximum Length and Weight of Selected Mammals 16
1.1.11 Size of Selected Breeds of Dog 18
1.1.12 Height at the Withers (= Shoulder Height) of Horses 19
1.1.13 Size and Weights of Selected Extinct Vertebrates 20
1.1.14 Size of Horns and Teeth 21
1.1.15 Number of Vertebrae of Selected Vertebrates 22
1.1.16 Number of Cervical (Neck) Vertebrae of Selected Vertebrates 22
1.1.17 Relative Skeletal Weight of Selected Vertebrates 23
1.1.18 Size of Selected Animal Cells 23
1.1.19 Water Content of Vertebrate Tissue 24
1.1.20 Amino Acid Composition of the Protein of Vertebrates 25
1.1.21 Effects of Animal Poisons (Not Including Those of Snakes) 25
1.1.22 Effects of Snake Venoms 27
1.1.23 Poisonous Effect of Selected Insecticides on Vertebrates 28

1.2   **Movement, Locomotion** 29
1.2.0   Numbers that Amaze 29
1.2.1   Diameter of Muscle Fibers 30
1.2.2   Twitch Times for Muscles 30
1.2.3   Contraction Times of Selected Muscles 31
1.2.4   Maximum Muscular Strength of Selected Muscles 31
1.2.5   Load-Bearing Record of Selected Muscles 32
1.2.6   Number of Wingbeats per Second in Insects 32
1.2.7   Surface Areas and Surface Loads of Insect Wings 33
1.2.8   Wingbeat Frequencies in Birds 34
1.2.9   Surface Areas and Surface Loads of Bird Wings 35
1.2.10 Maximum Speeds of Selected Animals 36

1.2.11   Jumping Distances of Selected Animals   38
1.2.12   Flight Speeds During Bird Migration   38
1.2.13   Flight Distances During Bird Migration   39
1.2.14   Migratory Times of Selected Birds Native to Germany   40

1.3   **Alimentation, Digestion**   41
1.3.0   Numbers that Amaze   41
1.3.1   Absorption of Nutrients in the Body   42
1.3.2   Required Daily Food Intake   42
1.3.3   Dental Formulae of Selected Mammals   43
1.3.4   Intestinal Length of Selected Vertebrates   46
1.3.5   Relative Intestinal Length of Selected Animals   47
1.3.6   Proportion of the Total Volume of the Intestinal Tract Made Up
         by Stomach, Small Intestine, and Appendix in Selected Mammals   48
1.3.7   Capacity of the Digestive Tract and its Parts
         in Selected Domestic Animals   48
1.3.8   pH Optimum of Selected Digestive Enzymes in Mammals   49
1.3.9   pH Value of the Saliva of Selected Mammals   50
1.3.10   pH Values in the Intestinal Tract of Selected Birds   50
1.3.11   pH Values in the Intestinal Tract of Selected Mammals   51
1.3.12   pH Values in the Mid-intestine of Selected Invertebrates   51
1.3.13   Duration of Intestinal Passage in Selected Vertebrates   52
1.3.14   Relative Weight of Liver in Selected Vertebrates   52
1.3.15   Daily Bile Production in Selected Vertebrates   53

1.4   **Metabolism**   54
1.4.1   Basal Metabolic Rate of Selected Animals   54
1.4.2   Body Temperatures of Selected Homoiothermic Animals   55
1.4.3   Size of Daily Variations in Body Temperature
         in Selected Homoiothermic Animals   57
1.4.4   Preferred Temperatures of Selected Animals   57
1.4.5   Duration of Hibernation (Torpor, Winter Rest)
         of Selected Vertebrates Native to Germany   59

1.5   **Excretion**   60
1.5.1   Number of Glomeruli in the Kidneys of Selected Vertebrates   60
1.5.2   Urine Production of Selected Mammals   61
1.5.3   Maximum Urine Concentration in Selected Mammals   61
1.5.4   Composition of the Excrement of Selected Vertebrates   62
1.5.5   Urea in the Urine of Selected Mammals   62
1.5.6   Freezing Point Depression and Osmotic Pressure
         of Urine in Selected Animals   63
1.5.7   Aquatic Salt Concentration Tolerance Limits of Selected Animals . . .   64

Contents

1.6	**Heart, Circulation**	65
1.6.0	Numbers that Amaze	65
1.6.1	Heart Weights of Selected Vertebrates (Absolute and Relative)	66
1.6.2	Heart Rates of Selected Animals	68
1.6.3	Heart Minute Volumes of Selected Animals	70
1.6.4	Blood Pressure of Selected Animals	71

1.7	**Blood**	72
1.7.1	Blood Volumes of Vertebrates	72
1.7.2	Number and Size of Red Blood Cells in Selected Vertebrates	73
1.7.3	Number of Platelets of Selected Animals	74
1.7.4	Coagulation Times of Blood in Vertebrates	74
1.7.5	Lifespan of Blood Cells	75
1.7.6	Hematocrit Levels of Selected Vertebrates	76
1.7.7	Hemoglobin Levels in the Blood of Selected Animals	76
1.7.8	Molecular Weight of the Hemoglobin of Selected Animals	77
1.7.9	Oxygen Capacity of the Blood (or Hemolymph) of Selected Animals	77
1.7.10	Blood Sugar Levels in Selected Vertebrates	79
1.7.11	Protein Levels of Bodily Fluids (Blood, Hemolymph, Coelom Fluid) of Selected Animals	80
1.7.12	Freezing Point Depression and Osmolarity of the Blood of Selected Animals	81
1.7.13	Freezing Point Depression und Osmotic Pressure of Hemolymphs of Arthropods	82
1.7.14	Physiological Salt Solutions (Ringer's, Tyrode's, Locke's Solutions) for Selected Animals	82

1.8	**Respiration**	83
1.8.0	Numbers that Amaze	83
1.8.1	Relative Lung Weight in Selected Vertebrates	84
1.8.2	Interior Lung Surface of Selected Animals	85
1.8.3	Breathing Rate and Single-Breath Air Volume of Selected Vertebrates	85
1.8.4	Oxygen Consumption of Selected Animals	87
1.8.5	Oxygen Consumption During Hibernation of Selected European Mammals	87
1.8.6	Maximum Diving Depth and Diving Time of Air-Breathing Vertebrates	88

1.9	**Reproduction, Development**	89
1.9.0	Numbers that Amaze	89
1.9.1	Size of Egg Cells in Selected Animals (Excluding Birds)	90
1.9.2	Dimensions of Sperm of Selected Animals	91
1.9.3	Chromosome Sets of Selected Animals	91

1.9.4   Age of Sexual Maturity in Selected Animals   93
1.9.5   Cell Division Time in Protozoa   94
1.9.6   Reproduction and Development of Selected Insects   94
1.9.7   Reproductive and Developmental Data
         of Selected Butterflies Native to Germany Over the Course of the Year .   95
1.9.8   Spawning Data of Selected Fish Native to Germany   96
1.9.9   Number and Size of Eggs per Clutch in Selected Fish   96
1.9.10   Number of Annual Offspring (Eggs, Larvae, Young)
         in Fish, Amphibians, and Reptiles   97
1.9.11   Reproductive Data of Amphibians and Reptiles Native to Germany . . .   98
1.9.12   Number of Eggs per Clutch in Selected Bird Species   99
1.9.13   Egg Size and Weight in Selected Bird Species   100
1.9.14   Length of Time of Incubation and Nesting
         (in Nidicolous Members) of Selected Bird Species   101
1.9.15   Incubation Dates of Selected, Well-Known Birds Native to Germany . .   102
1.9.16   Composition of the Chicken Egg   103
1.9.17   Embryonal Development of the Chick   104
1.9.18   Embryogenesis in Selected Mammals and Birds   104
1.9.19   Relative Length of Time of Sexual Maturation in Selected Mammals . .   105
1.9.20   Duration of Cycle and Estrus in Selected Mammals   106
1.9.21   Duration of Gravidity (Gestation Time)
         and Number of Young of Selected Mammals   106
1.9.22   Relative Birth Weight of Selected Mammals   108
1.9.23   Normal Growth (Mean Values of Body Weight)
         of Selected Pets and Laboratory Animals   108
1.9.24   Times for Doubling Birth Weight in Selected Mammals   109
1.9.25   Composition of Milk in Selected Mammals   110
1.9.26   Annual Milk Production of Domestic Animals   111
1.9.27   Lactation Periods of Selected Mammals   111
1.9.28   Closed Seasons for HunAnimals in Germany   112

1.10   Nervous System   113
1.10.0   Numbers that Amaze   113
1.10.1   Brain Weight (Absolute) of Selected Mammals   114
1.10.2   Relative Brain Weight of Selected Vertebrates   114
1.10.3   Brain Surface Area of Selected Mammals   115
1.10.4   Cerebralization Index of Birds and Mammals   115
1.10.5   Distribution of Important Ions on the Cell Membranes
         of Selected Excitable Structures   116
1.10.6   Resting Potentials of Nerve and Muscle Fibers   116
1.10.7   Nerve Conduction Velocity   117
1.10.8   Size of the Discharge of Electric Fish   117

1.11	**Senses**	118
1.11.1	Minimum Stimuli for Activating a Sensory Organ	118
1.11.2	Size of the Eyeball in Selected Animals	118
1.11.3	Focal Distance of the Lenses in the Eyes of Selected Animals	119
1.11.4	Number of Photoreceptors in the Eyes of Selected Animals	119
1.11.5	Number of Photoreceptors per Dendrite in the Eyes of Selected Vertebrates	120
1.11.6	Visual Acuity Angle (Angle of Minimum Separability) of Selected Animals	120
1.11.7	Visual Field (Horizontal) of Both Eyes in Selected Animals	121
1.11.8	Number of Unit Eyes (Ommatidia) in the Compound Eyes of Selected Arthropods	122
1.11.9	Flicker Fusion Rate of the Eyes of Selected Animals	122
1.11.10	Absorption Maxima of the Visual Pigments of Selected Invertebrates . .	123
1.11.11	Absorption Maxima of the Rhodopsins of Selected Vertebrates	123
1.11.12	Eardrum Surface Area of Selected Animals	124
1.11.13	Upper Limit of Hearing Ability in Selected Animals	124
1.11.14	Auditory Acuity of Selected Animals	125
1.11.15	Number of Taste Buds of Selected Vertebrates	126
1.11.16	Dimensions of Olfactory Epithelium in Selected Animals	126
1.11.17	Vibration Threshold at Optimal Frequency	127

Part 2: Botany

2.1	**General Data: Dimensions, Age, etc.**	131
2.1.0	Numbers that Amaze	131
2.1.1	Number of Species of Plant Taxa and Fungi	132
2.1.2	Number of Species Within the Individual Taxa of Seed-Bearing Plants .	132
2.1.3	Maximum Age of Selected Plants	133
2.1.4	Rotation Periods of Important Harvested Trees	134
2.1.5	Lifespan of Leaves and Needles of Selected Evergreen Plants	135
2.1.6	Root Depth of Selected Plants	135
2.1.7	Leaf Surface Areas of Selected Plants	136
2.1.8	Diameter of the Blossoms of Selected Plants	136
2.1.9	Specific Weight of Woods	137
2.1.10	Water Content of Selected Plant Organs	138

2.2	**Cellular Values**	139
2.2.1	Dimensions of Selected Plant Fibers	139
2.2.2	Vessel Diameters in Woods	139
2.2.3	Number of Stomata on the Leaves of Selected Plants	140
2.2.4	Length of the Stoma Pore in Selected Plants	141
2.2.5	Size of Chloroplasts of Selected Plants	142
2.2.6	Size of Starch Grains in Selected Useful Plants	143
2.2.7	Size of Pollen Grains in Selected Plants	143

2.2.8   Dimensions of Spores of a Selection of Well-Known Species   144
2.2.9   Chromosome Sets of Selected Plants   145

2.3   **Germination and Growth**   147
2.3.1   Germination Time and Optimal Germination Temperature
        of the Seeds of Selected Plants   147
2.3.2   Germination Speed of the Seed of Selected Useful Plants   148
2.3.3   Vernalization Temperatures and Periods of Selected Plants   148
2.3.4   Growth Speed of Selected Plant Organs Under Optimal Conditions . . .   149
2.3.5   Annual Vertical Growth Over Lifespan
        of Selected Important Harvested Trees Under Optimal Conditions . . .   150
2.3.6   Height of Selected Trees   150
2.3.7   Maximum Trunk Diameters of Selected Tree Species   151
2.3.8   Greatest Recorded Trunk Circumferences for Selected Trees   152
2.3.9   Slenderness Ratio and Height of Selected Plants   152
2.3.10 Length of Runners of Selected Plants Over the Course of a Year   153
2.3.11 Cardinal Temperature Points for the Growth of Selected Useful Plants .   154
2.3.12 Optimal pH Soil Levels for Plant Growth   155
2.3.13 Polar Limits for the Cultivation of Selected Useful Plants   156
2.3.14 Dryness Limits for the Cultivation of Selected Useful Plants   157

2.4   **Reproduction: Blossoms, Fruits, Seeds**   158
2.4.1   Linnaeus' Flower-Clock   158
2.4.2   Floral Formulas of Selected Plant Families   159
2.4.3   Blossoming Period of Selected Plants   161
2.4.4   Critical Day Length for Inducing Blossoming
        in Long-Day and Short-Day Plants   161
2.4.5   Blossoming Data for Selected Trees and Shrubs Native to Germany . . .   162
2.4.6   Blossoming Data for Selected Monocotyledonous Plants
        Native to Germany   164
2.4.7   Blossoming Data for Selected True Grasses (Poaceae)
        Native to Germany   165
2.4.8   Blossoming Data for Selected Dicotyledonous Plants Native to Germany   165
2.4.9   Time of Spore Maturation in Selected Ferns Native to Germany   168
2.4.10 Formation Time of Fruiting Bodies of Fungi Native to Germany   169
2.4.11 Annual Number of Seeds of Selected Plants   170
2.4.12 Number of Seeds per Capsule in Selected Plants   171
2.4.13 Weight of Selected Seeds   171
2.4.14 Thousand Seed Weights (TSW) und Hectoliter Weights
        of the Seeds of Selected Useful Plants   172
2.4.15 Age of Seeds of Selected Plants   173
2.4.16 Maximum Distance Seeds Catapulted by Ballistic Fruits   174

2.5 **Constituents** 175
2.5.1 Composition of Cereal Grains 175
2.5.2 Composition of Legumes 175
2.5.3 Fat Content of Selected Plant Organs 176
2.5.4 Fat Content of the Seeds and Fruits of Selected Plants
as a Percentage of Fresh Weight 177
2.5.5 Raw Protein Content of Selected Seeds and Fruits 178
2.5.6 Composition of Tubers 178
2.5.7 Sugar Content of the Nectar of Selected Plants 179
2.5.8 Energy Contents of Plant Material 179

2.6 **Physiological Values** 180
2.6.1 Speed of Plasma Flow in Selected Plant Organs 180
2.6.2 Osmotic Pressure in the Cell Sap of Selected Plant Leaves 180
2.6.3 Osmotic Values in Plants 181
2.6.4 Acidity in the Pressed Juice of Fruits and Vegetables 182
2.6.5 Temperature Tolerances of Plants 182
2.6.6 Annual Levels of Evaporation of Selected Plants 184
2.6.7 Transpiration Levels of Selected Plants 184
2.6.8 Cuticular Transpiration in Selected Plants 185
2.6.9 Water Consumption for Selected Plants Required for Production
of 1 Kg Plant Dry Matter (So-Called Transpiration Coefficient) 185
2.6.10 Hydraulic Conductivity of Selected Woods 186
2.6.11 Speed of Hydraulic Conductance in Selected Plants 186
2.6.12 Respiration Rates of Selected Lower Plants (Thallophyta) 187
2.6.13 Respiration Rates of Selected Plants and Plant Parts 187
2.6.14 Sugar and Starch Production in Photosynthesis 188
2.6.15 Annual Worldwide Photosynthesis Production and Carbon Exchange . 189
2.6.16 Speed of Rotation in Selected Twining Plants 190
2.6.17 Resting Potential of Plant Cells and Concentration
of Potassium Ions in the Membrane (Plasmalemma) 191
2.6.18 Conduction Velocity in Selected Plants 191
2.6.19 Absorption Maxima of Plant Pigments 192

Part 3: Microbiology

3.1 **Size and Structure of Viruses** 195
3.2 Dimensions of Bacteria 195
3.3 Generation Time of Selected Bacteria 196
3.4 pH Ranges for the Growth of Bacteria and Fungi 197
3.5 Lethal Temperatures for Selected Microorganisms 197
3.6 Inactivation Temperatures and Times for Selected Viruses 198
3.7 Respiration Rate of Selected Bacteria and Fungi 199
3.8 Incubation Times of Several Important Infectious Diseases 200

Part 4   Human Biology

4.1   General Data 203
4.1.0   Numbers that Amaze 203
4.1.1   Life Span of Cells in Selected Organs of the Human Body 204
4.1.2   DNA and RNA Contents of Selected Human Cells 204
4.1.3   Composition of the Human Body from the Elements 205
4.1.4   Composition of the Body of an Adult Human 205
4.1.5   Relative Share of Individual Tissues in Body Composition
         of Man and Woman 206
4.1.6   Growth of the Human Embryo During Pregnancy 206
4.1.7   Normal Growth of Boys and Girls 207
4.1.8   Number of Bones in the Human Body 208
4.1.9   Dimensions and Weight of Selected Human Organs 209
4.1.10   Specific Weight of Selected Human Organs 209
4.1.11   Water Content of Human Organs and Tissue 209
4.1.12   Temperature of Selected Organs and Parts of the Human Body 210
4.1.13   pH-Values of Human Bodily Fluids 211

4.2   Skin, Skin Adnexa (Hair, Nails, Glands) 212
4.2.1   Human Skin in Numbers 212
4.2.2   Number of Human Hairs 212
4.2.3   Number of Hairs per cm2 on Various Parts of the Body 213
4.2.4   Hair Diameters (Averages) in an Adult Male 213
4.2.5   Speed of Hair and Nail Growth 214
4.2.6   Number of Perspiratory Glands 214

4.3   Heart, Circulation, Blood 215
4.3.1   Numbers on the Heart and Heartbeat of Humans 215
4.3.2   Temporal Dimensions of the Human Heartbeat 215
4.3.3   Dimensions of the Human Blood Vessel System 216
4.3.4   Diameter of the Arteries of an Adult Human 217
4.3.5   Blood Pressure at Different Points in the Human Circulatory System .. 217
4.3.6   Age and Sex Dependency of Blood Pressure in Humans 217
4.3.7   Blood Flow Through Selected Human Organs 218
4.3.8   Proportional Blood Flow Through Selected Human Organs 219
4.3.9   Composition of Human Blood 219
4.3.10   Most Important Proteins in Human Blood Plasma 220
4.3.11   Quantity of Hemoglobin (Hb) and Number of Blood Cells in Humans . 221
4.3.12   Number of Leukocytes in Humans 221
4.3.13   Composition of Lymph and Blood Plasma in Humans – A Comparison 222
4.3.14   Percentual Distribution of Blood Groups in Selected Peoples 222

4.4 **Respiration** 223
4.4.2 Age and Sex Dependency of Vital Capacity in Humans 223
4.4.3 Average Air and Energy Consumption
of an Adult During Selected Activities 224
4.4.4 Changes in O_2 and CO_2 During Human Respiration 224
4.4.5 Age and Sex Dependency of Human Basal Metabolic Rate 224
4.4.6 Oxygen Consumption and Heat Generation in Human Organs 225

4.5 **Nutrition, Digestion, Excretion** 226
4.5.1 Standard Food Requirements for Adults 226
4.5.2 Recommended Daily Protein and Overall Energy
Intake in the Human Diet 226
4.5.3 Human Digestive Capacity for Selected Nutrients 227
4.5.4 Components of a Chemically Defined Diet (CDD)
that Meets Human Nutritional Needs 228
4.5.5 Energy Requirements in Humans of Selected Physical Activities 229
4.5.6 Composition and Energy Content of Selected Foodstuffs 230
4.5.7 Daily Human Vitamin Requirements 232
4.5.8 Vitamin Content of Selected Foodstuffs 233
4.5.9 Adult Daily Requirements of Essential Amino Acids 235
4.5.10 Fatty Acid and Cholesterol Content of Selected Foodstuffs 235
4.5.11 Numbers on the Human Digestive System 236
4.5.12 Teeth Eruption Times in Humans 237
4.5.13 Numbers on the Human Kidney 237
4.5.14 Filtration and Reabsorption of Selected Substances
in the Human Kidney 238
4.5.15 Daily Urine Volume in Humans at Different Ages 238

4.6 **Brain, Nerves, Senses** 239
4.6.1 Numbers on the Human Nervous System 239
4.6.2 Brain Volumes of Hominids 239
4.6.3 Brain Weights of Several Eminent Humans 240
4.6.4 Sleep Requirements and Proportion of REM Sleep
for Humans of Different Ages 240
4.6.5 Number of Receptors and Afferent Nerve Fibers of Human Senses . . . 241
4.6.6 Numbers on the Human Eye and Eyesight 241
4.6.7 Changes in the Binocular Horizontal Visual Field of Humans with Age . 242
4.6.8 Changes in the Near Point of the Human Eye with Age 242
4.6.9 Numbers on the Human Ear 243
4.6.10 Age Dependency of Upper Hearing Limits 243
4.6.11 Sound Levels of Selected Sounds 243
4.6.12 Spatial Discrimination Threshold of Selected Sites on the Human Body 245
4.6.13 Number and Distribution of Hot and Cold Spots in Human Skin 245
4.6.14 Odor Threshold in Humans for Selected Distinctive Substances 246
4.6.15 Taste Thresholds in Humans 246

Part 5: Comparative Data

5.1	Composition of Animal and Plant Organisms from the Elements	249
5.2	Water Content of Organisms	249
5.3	Size and Molecular Weight of Selected Important Biological Molecules .	250
5.4	Redox Potentials of Selected Organic Compounds	251
5.5	Dimensions of Cells and Organelles	251
5.6	Speed of Movement of Cells and Organelles	252
5.7	Times of Cell Division in Selected Cells	253
5.8	Dimensions of Cell Nuclei	253
5.9	Dimensions of the Mitochondria of Selected Organs and Organisms . .	254
5.10	Mitochondria: Cell Share	254
5.11	DNA Content of Diploid Cell Nuclei	255
5.12	Size of DNA in Selected Organisms	256
5.13	Base Ratios in the DNA of Selected Organisms	257
5.14	Differences in the Number of Amino Acids in Humans Versus Selected Organisms as Exemplified by Cytochrome c.	258

References 259

Subject Index 267

Name Index 275

Introduction to the 6th German Edition*

Teaching at schools, colleges, and universities often requires quick recourse to comparative data, whether for working out and presenting certain regularities or for comparing the figures gained from experiment and observation to those of the literature. The tables presented here were compiled to put an end to the time-consuming search for such data.

This compilation is primarily intended for those studying to become teachers and those already teaching at schools, colleges, and universities, for studies have shown pupils' questions are usually morphological in character, involving properties such as body size and weight. This volume aims to give teachers rapid access to such numbers. Much of the data, due to its nature, has long been known and is thus found primarily in older literature, which is not always easy to procure; the most recent of research findings, however, have also been incorporated.

The tables are comparative in character, for only in this way can regularities be recognized. Use of statistical data has been the exception rather than the rule, since they are often subject to change. The tables cannot be complete and were not intended to be. There was conscious effort to select those organisms native to our geographical region and commonly found in biology lessons and practical training. Where possible, the data have been supplemented by extreme values, and these have been marked by an exclamation point (!) in the tables. Furthermore, in this 6th German edition* – taking KUNSCH/KUNSCH: *Der Mensch in Zahlen* [Humans in Numbers], 2001 as its model – several of the sections are prefaced by "Numbers that Amaze" tables in order to attract the reader's attention and to induce him or her to compare and search. Even though these tables, given their specialized topics, are chiefly aimed at biologists, most of them offer the interested lay reader items of interest.

The majority of the tables require little explanation. For this reason, there was a basic decision to forgo the addition of more detailed elucidating texts. Unless there were reasons to organize the data according to size or a specific system, the organisms in the tables are listed alphabetically. This is specifically indicated in all tables where this is not the case.

The fact that this book has now gone into its 6th edition and has also appeared in Russian shows that its guiding conception is basically correct. Nevertheless, inconsistencies have been eliminated and data updated for each new edition. The 6th edition has once again been thoroughly revised and expanded.

Ludwigsburg, Germany, Spring 2002   Rainer Flindt

*The basis for the present translation [Translator's note]

Part 1: Zoology

As a point of comparison, data on humans have been included in the majority of tables without explicit mention of this in the table legends.

1.1 General: Size, Weight, Age ...

1.1.0: Numbers that Amaze

This table brings together some interesting data from Sect. 1.1. Subsequent tables provide greater detail and provide a context of comparable data for these figures; references to the literature are also found there.

Largest and heaviest vertebrate that has ever lived (and still does):	Blue whale	Length: 33 m Weight: 136,000 kg
Heaviest extinct terrestrial vertebrate:	*Brachiosaurus*	Length: 27 m Weight: 50,000 kg
Largest extinct mammal:	Fossil rhinoceros	Length: 8.5 m Weight: 18,000 kg
Smallest mammal:	Etruscan shrew	Length: 4 cm Weight: 2 g
Largest bird:	Ostrich	Weight: 150 kg
Largest extinct bird:	Madagascar's elephant bird	Weight: 400 kg
Bird with the greatest wingspan:	Albatross	Span: 320 cm
Bird with the greatest wing surface area:	Griffon vulture	10,450 cm^2
Smallest bird:	Little woodstar (hummingbird)	Weight: 1.6 g
Smallest living snake:	*Leptotyphlops* sp.	Length: 11 cm
Largest living snake:	Reticulated python	Length: up to 10 m
Smallest living reptile:	Lizard-toed gecko	Length: 4 cm Weight: 2 g
Largest living caudate:	Giant salamander	Length: 150 cm
Smallest living caudate:	Pygmy salamander	Length: 4 cm
Largest extinct amphibian:	*Palaeosiren*	Length: 15 m
Largest living frog:	Goliath frog	Length: 40 cm
Smallest living frog:	Almiqui (Cuban banana frog)	Length: 1 cm
Largest living fish:	Whale shark	Length: 18 m Weight: 10,000 kg
Smallest living fish:	Dwarf goby	Length: 11 mm Weight: 0.2 g
Largest extinct fish:	*Carcharodon* sp.	Length: 33 m Weight: 125,000 kg
Largest living insect:	Walking stick *Pharnacia*	Length: 33 cm
Living insect with largest wing surface area:	*Coscinocera hercules*	Area: 300 cm^2

Table 1.1.1: Approximate Numbers of Species in Selected Major Animal Taxa

For figures on plants, see Tables 2.1.1 and 2.1.2.
In all, 1.4 million animal species have been described.

Figures from BERTELSMANN 1977, *Die Roten Listen* 1997, KÄSTNER 1973, KRAUS and KUBITZKI 1982, *Lexikon der Biologie* 1983-87, MEYER 1964, *Naturschutz aktuell* 1977, RENNER 1984, STORCH and WELSCH 1991

Taxon	on Earth	in Germany
Protozoa	40,000	3,200
Sponges	5,000	28
Cnidaria	10,000	130
Flatworms	16,100	1,100
Roundworms	23,000	1,600
Mollusks	130,000	500
Snails	85,000	380
Clams	25,000	100
Cephalopods	600	15
Segmented worms (annelids)	17,000	1,900
Arthropods	> 1,000,000	32,500
Arachnids	68,000	2,300
Crustaceans	50,000	900
Insects	1,000,000	28,900
Dragonflies	4,700	80
Cockroaches	4,000	12
Locusts, grasshoppers	20,000	80
True bugs	73,000	1,800
Beetles	350,000	5,727
Bees and ants (hymenoptera)	110,000	10,000
Butterflies	120,000	3,000
Flies, gnats	120,000	8,000
Echinoderms	6,500	34
Vertebrates	46,500	476
Fish (incl. lampreys)	20,600	130
Amphibians	3,300	21
Reptiles	6,300	14
Birds	8,600	273
Mammals	3,700	98

Table 1.1.2: Maximum Age of Selected Animals

Not all data are reliably documented. Some animals only reach the given age in captivity, not in the wild. Figures in parentheses are for individual pets or zoo animals. Reliable data for birds in the wild, based on ring recovery data, are marked with an "r". All figures, unless otherwise noted, are in years. D days, W weeks, M months. For plant data, see Table 2.1.3.

Figures from ALTMANN and DITTMER 1972, BERTELSMANN 1979, DORST 1972, FRÄDRICH and FRÄDRICH 1973, KORSCHELT 1924, KRUMBIEGEL 1955, NOBLE 1954, OPPENHEIMER and PINCUSSEN 1930, PLATT 1976, SPECTOR 1956, TANNER 1980, WOOD 1982

Species	Maximum age		Species	Maximum age	
Invertebrates					
Auster	12		Starfish	>5	
Ascaris worm	5		Stony coral	>28	
Beadlet anemone	15		Tapeworm	35	
Centipede (*Lithobius*)	5–6		Termite queen	25	
Common bedbug	6	M	Terrestrial nematodes	2–3	W
Common periwinkle	20		Trichina		
Crayfish	20–30		(living in the intestine)	5	M
Cuttlefish (*Sepia*)	5		Tarantula	15	
Earthworm	10		Trichina (encysted)	30	
Earwig	7		Water flea	108	D
Edible snail (*Helix pomatia*)	> 18		Worker bee	6	W
Freshwater (pearl) mussels	100				
Freshwater snail	2–3		**Vertebrates**		
Fruit fly (*Drosophila*)	46	D	**Fish**		
Giant clam	60–100		Carp	70–100	
Golden ground beetle	5		Eel	88	
Housefly	76	D	Goldfish	41	
Lancelet	7	M	Guppy	5	
Leech	27		Herring	20	
Lobster	45		Lungfish	18	
Lower-order crustacea	6	M	Mackerel	20	
Octopus	2–3		Pike	60–70	
Oyster	12		Plaice	30	
Planarian	14	M	Salmon	13	
Pond mussel	9		Sea horse	5	
Praying mantis	8		Sturgeon	152	
Queen bee	5		Trout	18	
Rotifer	2–3	D	Whale shark	70	
Sea anemone (*Cereus*)	66				
Sea cucumber	>10		**Amphibians**		
Sea urchin	7		Axolotl	>24	
Spiders	20		Common toad	40	
Sponges	50		Fire salamander	43	

Species	Maximum age		Species	Maximum age	
Giant salamander	55		Hummingbird	8	
Smooth newt	28		Larks	8	r
Tree frog	22		Magpie	25	
			Ostrich	62	
Reptiles			Owls	60–70	
American alligator	66		Oystercatcher	29	
Anaconda	31		Pelican	50–60	
Boa	40		Penguin	26	
Cobra	28		Pheasant	27	
European pond turtle	120		Pigeons/doves	35	
Galapagos giant tortoise	150		Red-throated loon or diver	23	r
Lizards (*Lacerta*)	5-8		Robin	11	r
Rattlesnake	19		Stork	70–100	22r
Slowworm	33	(-54)	Swallows	16	r
Tortoise	137		Swan	30	(-100)
Tuatara	100		Winter wren	5	
Birds			**Mammals**		
Barn owl	14	r	Beaver	20–25	
Blackbird	18, 10	r	Baboon	35	
Buzzard (or Am.: hawk)	24	r	Bactrian camel	29	
Canary	24	(34)	Bison	30	
Chaffinch	29		Brown bear	47	
Chicken	30		Capuchin monkey	41	
Cockatoo	100		Cat	35	
Common raven	69		Cattle	20–25	(-49)
Common swift	21	r	Chamois	25–30	
Condor	65		Chimpanzee	>50	
Crane	62		Common or Harbor seal	26	
Crow	118	20r	Deer	30	
Cuckoo	40		Dogs	15–20	(-34)
Duck	25	20r	Dolphin	25–30	
Eagle	60–80	21r	Domestic pig	27	
Eagle owl	68		Donkey	100	
European starling	20	r	Dromedary	28	
Garden warbler	24		Elephant	70	
Goose	31	(50)	European hare	8	
Gray heron	60	24.5r	Fallow deer	25	
Great cormorant	21	18r	Fat dormouse	9	
Great crested grebe	23		Fox	14	
Great tit	9		Gibbon	>23	
Griffon vulture	118		Giraffe	34	
Herring gull	44	32r	Goat	20	
House sparrow	23		Golden or Syrian hamster	4	

(Cont.) →

Species	Maximum age	Species	Maximum age
Gorilla	60	Orangutan	59
Gray seal	42	Polar bear	41
Guinea pig	9.5 (-15)	Rabbit	18
Hedgehog	14	Rat	3
Hippopotamus	54	Reindeer	16
Horses	40–50 (-61)	Rhinoceros	45
Horseshoe bat	24	Roe deer	16
Human	118	Sea lion	28
Ibex	30	Sheep	20
Kangaroo	30	Shrews	1.5
Lion	30	Squirrel	12
Llama	15	Tiger	30
Long-beaked echidna	49	Whale	100
Mole	3–4	Wild boar	20–30
Moose (in Europe: elk)	25	Wolf	14
Mouse	4	Wood mouse	10 M
Mule	45	Zebra	38
Myotis (bat)	24		

Table 1.1.3: Largest Representatives of Selected Animal Taxa (Excluding Vertebrates and Insects)

For data on vertebrates, see Tables 1.1.6 through 1.1.12; for insects, Table 1.1.4.
L = length, D = diameter, W = width, S = span.

Figures from FRANK 1976, KÄSTNER 1973

Taxon	Species	Dimension
Protozoa	*Nummulites*	D: 32 cm
Sponges	*Spheciospongia*	D: 2 m
Cnidaria		
Hydrozoa	*Brachiocerianthus imperialis*, Polyp	L: 2.20 m
	Aequorea sp., jellyfish	D: 40 cm
Scyphozoa	*Cyanea arctica*, jellyfish	D: 2 m
Sea anemone	*Stoichactis* sp.	D: 1.50 m
Solitary corals	*Fungia* sp.	D: 25 cm
Ctenophores	*Cestus veneris*	L: 1.50 m
Flatworms		
Turbellaria	Terrestrial planarian	L: 60 cm
Trematodes	*Nematobothrioides histoidii*	L: 6-8 m
Tapeworms	Diphyllobothrium latum	L: 15 m

Taxon	Species	Dimension
Ribbon worms	*Lineus longissimus*	L: 30 m
Rotifers	*Seison* sp.	L: 3 mm
Roundworms	*Placentonema gigantissima* ♂	L: up to 3.75 m
	♀	L: up to 8.40 m
Mollusks		
Snails	*Semifusus* sp.	L: 60 cm (shell)
Clams	*Tridacna gigas*	L: 1.35 m (shell)
Cephalopods	*Architeuthis* sp.	L: 6.60 m (without tentacles)
		18 m (with tentacles)
Segmented worms (annelids)		
Polychaetes ("many bristled")	*Eunice gigantea*	L: 3 m
Oligochaetes ("few bristled")	*Megascolides australis*	L: 3 m
Leeches	*Haementeria ghiliani*	L: 30 cm
Arthropods		
Scorpions	*Pandinus imperator*	L: 18 cm
Spiders	*Teraphosa leblondi* (tarantula)	L: 9 cm (torso)
Crustaceans	*Janus hügeli*	L: 60 cm (without legs)
	Macrocheira kaempferi	W: 29 cm (carapace)
		S: 3 m (legs)
Myriapoda	*Scolopendra gigantea* (centipede)	L: 26.5 cm
	Graphidostreptus gigas (diplopod)	L: 28 cm
Ectoprocta (moss animals)		*Nolella alta* L: 4.6 mm
(single animal)		
	Pectinatella gelatinosa	L: 2 m (colony)
Acorn worms	*Balanoglosssus gigas*	L: 2.50 m
Echinoderms		
Crinoids (feather-stars)	*Metacrinus superbus*	L: 19 cm (arms)
		2 m (stem)
	Heliometra glacialis	L: 35 cm (arms)
Sea cucumbers	*Synapta maculata*	L: 2 m
		D: 5 cm
	Stichopus variegatus	L: 1 m
		D: 21 cm
Sea urchin	*Sperosoma giganteum*	D: 32 cm
Starfish	*Freyella remex*	L: 45 cm
		S: 1 m
Brittle stars	*Gorgonocephalus stimpsoni*	L: 70 cm (arms)
		D: 14.3 cm (disk)
Arrow worms	Sagitta gazellae	L: 10 cm

Table 1.1.4: The Largest Insects

L = length, WS = wingspan

Data after Kästner 1973

Taxon	Species	Size				
Mayflies	*Euthyplocia* sp.	L: 5 cm,		WS: 8 cm		
Dragonflies	*Megaloprepus coerulatus*	L: 15 cm,		WS: 18 cm		
Stoneflies	*Diamphipnoa helgae*	L: 4 cm,		WS: 11 cm		
Earwigs	*Titanolabis colossa*	L: 5 cm				
Praying mantis	*Toxodera* sp.	L: 16 cm				
Cockroaches	*Megaloblatta longipennis*	L: 9.5 cm,		WS: 17 cm		
Termites	*Macrotermes goliath*	L: 2.2 cm,		WS: 8.8 cm		
	Bellicositermes natalensis ♂	L: 14 cm				
Stick insects (Phasmida)	*Pharnacia serratipes*	L: 33 cm				
Saltatoria						
Katydids (bush crickets)	*Saga syriaca*	L: 10 cm				
	Pseudophyllus colossus			WS: 20 cm		
Grasshoppers	*Tropidacris latreillei*	L: 12 cm,		WS: 23 cm		
True bugs						
Heteroptera	*Belostoma grande*	L: 11 cm				
Cicadas	*Pomponia imperatoria*	L: 7 cm,		WS: 18 cm		
Megaloptera	*Sialis flavilatera*			WS: 3.8 cm		
Snakefly	*Acanthocorydalus kolbei*			WS: 16 cm		
Beetles	*Titanus giganteus*	L: 15 cm				
Bees and ants (hymenoptera)	*Pepsis hero*	L: 6 cm,		WS: 11 cm		
Neuroptera	*Palpares voeltzkowi*			WS: 15.7 cm		
Butterflies	*Thysania agrippina*			WS: 32 cm		
	Coscinocera hercules	wing surface: 300 cm^2				
Diptera	*Mydas hero*	L: 6 cm,		WS: 10 cm		
Fleas	*Histrichopsylla talpae*	L: 6 mm				

Table 1.1.5: Size of Populations in Colony-Building Insects

Listed according to size, in descending order.

Figures from GRZIMEK 1970, JAKOBS and RENNER 1974, KÄSTNER 1973, KEMPER and DÖHRING 1967, MEYER 1964, WEBER 1966

Species	Size of Population
Tropical termites (*Bellicositermes*)	3,000,000
Southern wood ant (*Formica rufa*)	500,000–800,000
Leafcutter ant (*Atta*)	600,000
Driver ants (*Anomma*)	100,000
Western subterranean termites (*Reticulitermes*)	100,000
Honeybees	40,000–80,000
Yellow meadow ant (*Lasius flavus*)	20,000
German yellowjacket wasp (*Paravespula germanica*)	5,500
Hornets	700–1,500
Bumblebees	500–2,000
Dolichovespula wasps	200
Paper wasp (*Polistes*)	140
Ant (Promyrmecia aberrans)	12

Table 1.1.6: Size and Weight of Selected Fish

The maximum sizes and weights reported here represent, in some cases, one-time record catches. For some species, the world-record sizes and weights for fish caught by angling are given in parentheses. The lengths represent total length.

Data after BLINKER 2001, GOADBY 1996, GRZIMEK 1970, LADIGES and VOGT 1965, MEYER 1964, NORMAN and FRASER 1963, SLIJPER, SPECTOR 1956, WOOD 1982

Species	Normal length	Maximum length (Fishing record)	Weight, up to (Fishing record)
Arapaima	2.5 m	4.5 m	400 kg
Barracuda	1.5–1.8 m	3.0 m (1.68)	50 kg (46.83)
Basking shark		14.0 m	4,000 kg
Beluga	4.0 m	9.0 m	1,600 kg
Black marlin		4.5 m (4.42)	900 kg (707.6)
Blue shark	3.0–4.0 m	6.0 m	800 kg
Carp	0.4–0.5 m	1.2 m (1.07)	30 kg (30.1)
Cod	1.0–1.5 m	1.8 m	95 kg
Dwarf goby (*Pandaka pygmaea*; smallest fish and smallest vertebrate)		11 mm (!)	0.2g (!)
Eel ♀		150 cm	6 kg (6.04)
Eel ♂		50 cm	1 kg
European conger	1.5–1.8 m	3.0 m	70 kg
European perch	0.3 m	0.45 m	2.56 kg
Greenland shark	3.0–4.0 m	7.0 m	
Grouper	2.0–2.2 m	3.5 m	420 kg (248.6)
Halibut	2.0–2.5 m	4.7 m	330 kg
Latimeria (Coelacanth)		1.5 m	52 kg
Lungfish (Australian)	0.6–1.2 m	1.8 m	50 kg
Man-eating shark	5.0–6.0 m	12.0 m (4.60)	3,000 kg (1,075.9)
Manta ray		4.0 m	1,600 kg
		7.0 m ("wing" span)	
Northern bluefin tuna	3.0 m	5.0 m (2.95)	820 kg (443.2)
Ocean sunfish		3.0 m	900 kg
Pike 1.0 m	1.5 m	(1.33) 70 kg	(25)
Salmon	0.9–1.0 m	1.5 m	36 kg (35.89)
Sawfish	6.0 m	10.0 m	2,300 kg
Sea lamprey		1.0 m	1 kg
Swordfish	3.0–4.0 m	6.0 m (4.55)	600 kg (681.8)
Tench	0.4–0.50 m	0.70 m	2 kg (4.97)
Tiger shark	4.0–5.0 m	6.0 m (4.22)	800 kg (626.9)
Trout	0.3 m	0.50 m	16 kg (15.7)
Wels catfish	2.0–2.5 m	5.0 m	300 kg (82)
Whale shark	15.0 m	18.0 m (!)	10,000 kg (!)
Zander		1.2 m	18 kg (18.6)

Table 1.1.7: Maximum Length of Selected Amphibians

For the salamanders (caudates) and caecilians, figures refer to total length; for the anura, they refer to crown–rump length (CRL). For figures on extinct amphibians, see Table 1.1.13.

Figures from COCHRAN 1970, GRZIMEK 1970, MERTENS 1947, MEYER 1964

Species	Length	
Caecilians and salamanders (caudates)		
Alpine newt ♀	11 cm	
Alpine newt ♂	8 cm	
Amphiuma ("Congo eel")	100 cm	
Axolotl	33 cm	
Caecilians	150 cm	
Fire salamander	20 cm	
Giant salamander	150 cm	(!)
Greater siren	100 cm	
Mexican pygmy salamander	4 cm	(!)
Northern crested newt	14 cm	
Palmate newt ♀	9.2 cm	
Palmate newt ♂	7.5 cm	
Smooth newt ♀	9.5 cm	
Smooth newt ♂	11.1 cm	
Anura		
Agile frog	8 cm	
Almiqui (Cuban banana frog)	1 cm	(!)
Colombian giant toad	30 cm	
Common frog	10 cm	
Common spadefoot toad ♀	8 cm	
Common spadefoot toad ♂	6.5 cm	
Common toad ♀	13 cm	
Common toad ♂	8 cm	
Cuban tree frog	13 cm	
Giant neotropical toad	20 cm	
Goliath frog	40 cm	(!)
Green toad	8 cm	
Marsh or Lake frog	15 cm	
Midwife toad	5 cm	
Moor frog	6.5 cm	
Natterjack toad	7 cm	
Tree frog	5 cm	
Yellow-bellied toad	4.5 cm	

Table 1.1.8: Maximum Length and Weight of Selected Reptiles

The figures for turtles refer to shell length; for all other animals, they refer to total length. Figures for extinct species can be found in Table 1.1.13.

Data after BELLAIRS 1971, GRZIMEK 1970, MANN 1979, MERTENS 1947, SCHMIDT and INGER 1957, WOOD 1982

Species	Length (m)		Weight
Turtles			
Alligator snapping turtle	0.75		100 kg
Green turtle	1.40		200 kg
Leatherback turtle	3.00		800 kg
Seychelles giant tortoise	1.25		250 kg
Crocodilians			
American alligator	5.84		
Gharial	7.11		
Nile crocodile	6.70		1,000 kg
Orinoco crocodile	7.20		
Saltwater or estuarine crocodile	8.50		
Lizards			
Common lizard	0.16		
Common wall lizard	0.19		
Dwarf gecko (*Jaragua sphaero*)	0.016	(!)	
Eyed or Jeweled lizard	0.80		
Green iguana	2.20		
Komodo dragon	3.00		165 kg
Lizard-toed gecko	0.04		2 g
Marine iguana	1.75		
Sand lizard	0.32		
Slowworm	0.45		
Tuatara ♂	0.29		1 kg
Western green lizard	0.45		
Snakes			
Aesculapian snake	1.80		
Anaconda	9.60		120 kg
Black mamba	4.00		
Boa constrictor	4.50		60 kg
Bushmaster	3.75		
Crossed viper	0.80		
Dice snake	1.00		
Grass snake	1.50		
King cobra	5.58		

Species	Length (m)		Weight
Leptotyphlops bilineata (smallest snake)	0.11	(!)	
Reticulated python	10.00	(!)	
Smooth snake	0.75		
Western diamondback rattlesnake	2.58		10 kg

Table 1.1.9: Weight and Wingspan of Selected Birds

For wing surface area and surface loads, see Table 1.2.9. For figures on extinct birds, see Table 1.1.13.

Data after BERNDT and MEISE 1959, GRZIMEK 1970, HEINROTH 1955, HERZOG 1968, STURKIE 1976, ZISWILER 1976

Species	Weight, up to			Wingspan, up to	
Albatross	8	kg		320 cm	(!)
Barn owl	300	g		95 cm	
Black woodpecker	315	g		75 cm	
Blue tit	12	g		20 cm	
Carrion crow	600	g		100 cm	
Common buzzard	1.2	kg		140 cm	
Common raven	1,250	g		125 cm	
Common swift	43	g		35 cm	
Condor	11.3	kg		290 cm	
Crest or Kinglet	4	g		16 cm	
Eagle owl	3.2	kg		170 cm	
Eurasian black vulture (Monk vulture)	14	kg		287 cm	
Eurasian curlew	600	g		100 cm	
Eurasian jay	195	g		55 cm	
Eurasian pygmy owl	60	g		40 cm	
Eurasian woodcock	400	g		58 cm	
Golden eagle	6	kg		230 cm	
Goshawk	2	kg		118 cm	
Great black-backed gull	2.2	kg		175 cm	
Great bustard ♂	5	kg			
Great bustard ♀	22	kg		240 cm	
Great spotted woodpecker	95	g		48 cm	
Griffon vulture	8.2	kg		240 cm	
Imperial eagle	3.5	kg		215 cm	
Little woodstar (smallest hummingbird)	1.6	g	(!)	7.5 cm	(!)
Marabou stork	5	kg		300 cm	
Mute swan	22.5	kg		260 cm	
Northern gannet	3	kg		180 cm	

(Cont.) →

Species	Weight, up to		Wingspan, up to
Northern lapwing	200	g	70 cm
Ostrich	150	kg (!)	
Painted quail	45	g	
Rock dove or pigeon	300	g	63 cm
Rook	670	g	90 cm
Sparrowhawk	300	g	76 cm
Stork	4.4	kg	220 cm
Tawny owl	500	g	100 cm
White-tailed eagle	6.7	kg	255 cm
Wood grouse or Capercaillie	6	kg	130 cm
Wood pigeon	500	g	75 cm

Table 1.1.10: Maximum Length and Weight of Selected Mammals

Figures on length refer to crown–rump length (CRL), except in the case of whales and pinnipeds, where they refer to total length.

Figures after BRINK 1975, FRÄDRICH and FRÄDRICH 1973, GRZIMEK 1970, KRUMMBIEGEL 1953, SANDERSON 1956, SLIJPER 1967, WOOD 1982

Species	Length, up to (cm)	Weight, up to	
Baboon	114	54	kg
Beaver	100	30	kg
Bison	300	1,000	kg
Black rhinoceros	375	2,000	kg
Blue whale	3,300 (!)	136,000	kg (!)
Bowhead whale	2,100		
Brown rat	26	400	g
Camel	345	650	kg
Chamois	130	62	kg
Chimpanzee	170 (standing height)	70	kg
Common eland	345	1,000	kg
Common house mouse	12	30	g
Common or Harbor seal	200	100	kg
Common vole	12	50	g
Dolphin	360	200	kg
Dwarf lemur	13	60	g
Elephant seal	650	3,600	kg
Elephant, African	750	6,000	kg
Elephant, Asian	640	5,000	kg
Ermine	29	450	g
Etruscan shrew	4 (!)	2	g (!)

Species	Length, up to (cm)		Weight, up to	
Eurasian badger	85		20	kg
Fat dormouse	19		120	g
Fin whale	2,500		76,000	kg
Fox	90		10	kg
Gibbon	90		22.5	kg
Giraffe	580	(height)	1,200	kg
Golden or Syrian hamster	18		130	g
Gorilla	175	(standing height)	275	kg
Grizzly bear	300		1,200	kg
Guinea pig	33		1,400	g
Hamster	34		500	g
Hazel dormouse	9		40	g
Hedgehog	30		1,200	g
Hippopotamus	450		3,200	kg
Horse (draft horse, "coldblood")			1,000	kg
Horse (warmblood)			500	kg
Ibex	170		150	kg
Indian rhinoceros	420		2,000	kg
Koala	80		16	kg
Least weasel	23		130	g
Leopard	150		80	kg
Lion	190		250	kg
Marmot	73		8	kg
Mole	17		120	g
Moose (in Europe: elk)	310		800	kg
Muskrat	36		1,500	g
Orangutan	150	(standing height)	100	kg
Orca	950		6,350	kg
Polar bear	251		1,000	kg
Pygmy mouse	6.6		6	g
Rabbit	45		2	kg
Red deer (Wapiti)	265		340	kg
Red kangaroo	160		70	kg
Reindeer	220		315	kg
Roe deer	140		50	kg
Sloth	60		5	kg
Sperm whale	2,000		53,000	kg
Square-lipped rhinoceros	400		3,600	kg
Squirrel	25		480	g
Tiger	300		350	kg
Walrus	375		1,500	kg
Water buffalo	300		1,000	kg
Water vole	19		180	g
Wild boar	180		350	kg
Wolf	150		75	kg
Wolverine (Glutton)	87		35	kg

Table 1.1.11: Size of Selected Breeds of Dog

Figures give crown–rump length (CRL) and height at withers (= shoulder height), both in centimeters.

Data after HALTENORTH 1958, SMOLIK 1968

Species	CRL		Withers	
Wolf	100–140		65–90	
Dingo	100		50–55	
Affenpinscher	20–25		26	
Airedale terrier	55–62		55–62	
Appenzeller mountain dog	49–58		48–58	
Barzoi	75–80		71–76	
Bloodhound	65–72		60–70	
Boxer	57–73		53–63	
Chihuahua	15–20	(!)	15–23	(!)
Cocker spaniel	35–42		35–42	
Collie	56–63		50–60	
Dachshund	23–30		23–27	
Dalmatian	50–60		50–60	
Fox terrier	35–40		35–60	
German shepherd	60–65		55–65	
Giant schnauzer (Riesenschnauzer)	57–70		60–70	
Great Dane	75–90	(!)	72–92	(!)
Greyhound	65		65–70	
Italian Greyhound	36		35	
Miniature poodle	35		30–35	
Newfoundland	63–75		62–75	
Pointer	60–65		60–65	
Pomeranian	28		28	
Poodle	45–60		45–55	
Pug	25–32		32	
Rottweiler	60–65		55–66	
St. Bernard	68–78		65–80	

Table 1.1.12: Height at the Withers (= Shoulder Height) of Horses

Figures from GRZIMEK 1970, KRUMMBIEGEL 1958, McGREGOR and EDWARDS 1976

Species	Height at the withers
Wild horses	
Grant's zebra	1.40 m
Grevy's zebra	1.55 m
Hartmann's zebra	1.25 m
Kiang	1.30 m
Kulan	1.70 m (!)
Onager	1.10 m
Przewalski's horse	1.20–1.46 m
Quagga	1.35 m
Somali wild ass	1.40 m
Horse breeds	
Andalusian horse	1.54 m
Anglo-Arabian	1.63–1.65 m
Arabian	1.44–1.50 m
Camargue	1.35–1.45 m
Clydesdale ("coldblood")	1.73 m
Exmoor pony	1.24 m
Fjord horse	1.40 m
Haflinger	1.21–1.48 m
Indian ponies	1.30–1.40 m
Lippizaner	1.55 m
New forest pony	1.22–1.44 m
Norwegian fjord horse	1.40 m
Percheron ("coldblood")	1.73 m
Ponies (greatest height)	1.44 m
Shetland ponies	0.95–1.05 m
Shire horse ("coldblood")	1.82 m (!)
Suffolk ("coldblood")	1.63 m

Table 1.1.13: Size and Weights of Selected Extinct Vertebrates

Figures from GRZIMEK 1970, KUHN 1958, KUHN-SCHNYDER 1974, LÜTTSCHWAGER 1961, MEYER 1974, SLIJPER 1967, WOOD 1982

Species	Length (m)	Height (m)	Estimated weight (kg)
Sharks			
Carcharodon megalodon	33		125,000
Fish (placodermi)			
Titanichthys	8		
Amphibians			
Mastodonsaurus	3		
Palaeosiren beinerti (Caecilian)	15		
Prionosuchus plummeri	9		
Reptiles			
Dinosaurs			
Brachiosaurus	27	15	50,000
Brontosaurus	22		40,000
Diplodocus	28		40,000
Iguanodon	11	5	10,000
Stegosaurus	9		1,750
Tyrannosaurus	16	6	
Crocodiles			
Phobosuchus	13.5		
Rhamphosuchus indicus	18.3		
Pterosaurs			
Pteranodon	wingspan 7 m		30
Quetzalcoatlus northropi	wingspan 12 m		86
Ichthyosaurs			
Ichthyosaurus	12		
Snakes			
Pterosphenus	11		
Turtles			
Archelon ischirus	6		2,700
Colossochelys atlas	6		
Birds			
Aepyornis (Madagascar's elephant bird)		1.5	400
Archaeopteryx	0.45		
Argentavis magnificens (capable of flying, wingspan 7.6 m)		1.5	120
Dodo		1	25
Giant moa		3.5	250

Species	Length (m)	Height (m)	Estimated weight (kg)
Mammals			
Baluchitherium (fossil rhinoceros)	8.5	5.5	18,000
Dinotherium (fossil elephant)		4	7,500
Cave bear	2.8		
Mammoth	3.5	4	7,500
Aurochs	3	1.8	1,500

Table 1.1.14: Size of Horns and Teeth

The figures represent maximum values from individual hunting trophies. Winding horns and teeth measured along their external curvature.

Data after Dorst and Dandelot 1973, Haltenorth and Trense 1956

Species	Size	
Antlers, horns		
Alpine ibex	147.32 cm	
Aurochs	80.00 cm	
Common eland	112.40 cm	
Impala	92.08 cm	
African buffalo	163.80 cm	
Greater kudu	168.91 cm	
Sable antelope	163.53 cm	
Gemsbok	121.92 cm	
Red deer (Wapiti)	152.40 cm	
Water buffalo	197.49 cm	
Black wildebeest	67.95 cm	
Frontal horn (rhinoceros)		
Square-lipped rhinoceros	158.12 cm	
Indian rhinoceros	60.96 cm	
Black rhinoceros	135.90 cm	
Teeth		
African elephant, tusk	349.25 cm	133.02 kg
Asian elephant, tusk	266.70 cm	72.98 kg
Hippopotamus, lower canine	163.83 cm	
Warthog, tusk	60.96 cm	

Table 1.1.15: Number of Vertebrae of Selected Vertebrates

Figures for neck vertebrae alone are found in Table 1.1.16.

After BOGNER and RITTER 1976, GRZIMEK 1970, MEYER 1964, OPPENHEIMER and PINCUSSEN 1930

Species	Vertebrae (n)	
Sharks	ca. 400	
Rainbow trout	60–69	
Herring	ca. 55	
Eel	max. 119	
Northern crested newt	53	
Frog	9	
Turtle	34	
Slowworm	110	
Grass snake	230	
Boas	max. 435	
Archaeophis (fossil boa)	565	(!)
Songbirds	35	
Swan	56	
Pets (mammals)	50–60	
Human	34	

Table 1.1.16: Number of Cervical (Neck) Vertebrae of Selected Vertebrates

For overall number of vertebrae, see Table 1.1.15.

Figures from GRZIMEK 1970, KUHN 1958, ZISWILER 1976

Taxon, species		Cervical vertrebrae (n)	
Amphibians		1	
Crocodiles		2	
Turtles		8	
Monitor lizards		8	
Heloderma		9	
Elasmosaurus (dinosaur)		76	
Birds, normally		14	(–15)
Exceptions:	Carrion crow	10	
	Albatross	15	
	Flamingo	19	
	Mute swan	26	
Mammals		7	
Exceptions:	Three-toed sloth (Genus *Bradypus*)	9	
	Hoffmann's two-toed sloth (*Choloepus hoffmanni*)	6	
	Manatees	6	

Table 1.1.17: Relative Skeletal Weight of Selected Vertebrates

Figures as percentage of body weight. Birds and mammals each listed according to percentage. Larger animals require a much sturdier skeleton than do their smaller counterparts.

Data after HESSE and DOFLEIN 1935, NACHTIGALL 1979, SLIJPER 1967

Species	Skeletal weight (%)
Birds	
Goose	13.3
Chicken	11.7
Sparrow	8.4
Winter wren	7.1
Mammals	
Elephant	25.0
Hippopotamus	20.0
Cattle	20.0
Sheep	20.0
Pig	18.0
Human	15.0
Dog	14.0
Lion	13.0
Cat	11.5
Common or Harbor seal	11.0
Rabbit	9.0
Common house mouse	8.4
Shrew	7.9

Table 1.1.18: Size of Selected Animal Cells

Figures for selected species and organs.

Data from DAVID 1977

Species	Organ	Diameter (μm)	Volume (μm^3)
Dog	Stomach	15.6	1,988
	Purkinje cells	30.0	14,000
Cat	Pyramidal cells		24,112
	Spinal ganglion		26,767
Horse	Pyramidal cells of Betz		9,348
	Purkinje cells	32.0	16,000

(Cont.) →

Species	Organ	Diameter (μm)	Volume (μm³)
Rabbit	Cerebral cortex		5,000–20,000
	Pyramidal cells of Betz		3,239
	Spinal ganglion		23,995
Mouse	Purkinje cells	19.0	4,000
	Spinal ganglion		4,908
Rat	Purkinje cells	16.0	2,000
	Spinal ganglion		14,440
Cattle	Purkinje cells	32.0	17,000
	Spinal ganglion		125,317
Pig	Purkinje cells	29.0	13,000
Frog	Purkinje cells	7.0	1,500
Chicken	Lymphocytes	8.0–13.0	
Macaque	Spinal ganglion		23,018
Human	Pyramidal cells of Betz		113,400
	Purkinje cells	27.0	10,000
	Muscle (M. psoas)	25.0–90.0	
	Erythrocytes		87
	Spinal ganglion		102,612

Table 1.1.19: Water Content of Vertebrate Tissue

See also Table 4.1.11 for more detailed data on human tissue.

From BUDDENBROCK 1956, modified

Species	Smooth muscles	Skeletal muscles	Skin	Kidney	Liver	Bones
Frog	85.0	82.1	81.1	84.4	76.1	58.1
Desert monitor	78.8	78.7	61.8	81.1	75.4	10.6
Rock dove	71.0	77.5	63.0	78.7	74.8	11.2
Gerbil	72.5	74.1	39.3	77.8	70.1	27.1
Lesser Egyptian jerboa	76.6	74.8	48.7	77.7	69.9	26.8
Human	80.0	76.8	70.0	82.5	76.9	30.0

Table 1.1.20: Amino Acid Composition of the Protein of Vertebrates

Figures represent percentage of weight.

From BUDDENBROCK 1956, modified

Amino acid	(%)
Glutamic acid	11.2–13.9
Glycine	8.2–10.2
Aspartic acid	8.1–10.4
Leucine	6.1– 7.5
Arginine	5.8– 6.3
Lysine	5.6– 7.9
Valine	4.4– 5.7
Threonine	3.7– 4.3
Isoleucine	3.3– 4.7
Phenylalanine	3.2– 4.2
Methionine	1.7– 2.1
Histidine	1.6– 2.1
Tryptophan	0.6– 1.0

Table 1.1.21: Effects of Animal Poisons (Not Including Those of Snakes)

Figures provided represent either the dose that kills half (50%) of the animals tested (LD_{50}, where LD = lethal dose) or the minimal lethal dose (MLD), calculated in terms of 1 kg body weight of the examined animal species (species in parentheses). Data on snake poison is found in Table 1.1.22.

Figures from BASSUS 1965, BERTELSMANN 1979, MEBS 1979

Animal species	MLD	LD_{50}
Cnidaria		
Congestin (from nematocysts of sea anemone)	5 mg (dog)	
Hypsotoxin (from nematocysts of the siphonophore *Physalia*)	2.0–2.5 mg (mouse)	
Lophotoxin (from nematocysts of the gorgonian coral *Lophogorgia*)	8 mg (mouse)	
Palytoxin (from nematocysts of the sea mat *Palythoa*)		0.01 µg (mouse)
Thalassin (from nematocysts of sea anemone)	2.5–9 mg (dog)	

(Cont.) →

Animal species	MLD			LD$_{50}$		
Mollusks						
Murex (poison of murex mollusk)	8.5	mg	(mouse)			
Arthropods						
Scorpion (*Buthus acutecarinatus*)	0.27	mg	(rat)			
Scorpion (*Tityus bahiensis*)	0.5–2.0	mg	(mouse)			
Spider (*Lycosa raptoria*)	0.9	mg	(mouse)			
Spider (*Ctenus ferus*)				0.34	mg	(mouse)
Tarantula (*Atrax robustus*)				4.5	mg	(mouse)
Cantharidin (from blister beetles)	0.5	mg	(human)			
Hornet venom				2.5	mg	(mouse)
Bee venom				3.0–4.0	mg	(mouse)
Poison of North American harvester ant (*Pogonomyrmex*)				0.42	mg	(mouse)
Fish						
Poison of stonefish				0.2	mg	(mouse)
Poison of pufferfish (Tetrodotoxin)	0.014–0.09	mg	(mouse)			
Amphibians						
Salamandrin (from salamanders)	0.1	mg	(rabbit)			
Skin poison from newts	0.018–0.02	mg	(mouse)			
Bufogenin (from toad skin)				0.13	mg	(cat)
Bufotenin (from toad skin)				25	mg	(mouse)
Poison from water frog (or pool frog) skin	6–12	mg	(rabbit)			
Dart poison (from *Phyllobates bicolor*)				2.7	mg	(mouse)
Batrachotoxin (from *P. aurotaenia*)				2.0	μg	(mouse)
Homobatrachotoxin (from *P. aurotaenia*)				3.0	μg	(mouse)
Pumiliotoxin (from *Dendrobates pumilio*)				2.0	mg	(mouse)
Reptiles (excluding snakes)						
Poison from Heloderma	16	mg	(mouse)			

Table 1.1.22: Effects of Snake Venoms

Figures provided represent either the dose that kills half (50%) of the mice injected with it
(LD_{50}, where LD = lethal dose) or the minimal lethal dose (MLD), calculated in terms of
1 kg body weight of the examined animal species (species in parentheses).
For the effect of the poison of other animals, see Table 1.1.21.

Data after BASSUS 1965, ZISWILER 1965

Animal species	Dry weight (mg) of poison per bite	LD_{50} (mg/kg) mouse	MLD (mg/kg) mouse/rabbit
Crossed viper			1–1.4 (mouse)
Russell's viper (*Vipera russeli*)	130–250	0.82	
Asp viper			1–2 (mouse)
Puff adder (*Bitis arietans*)	130–250	3.68	
Rattlesnake (*Crotalus durissus*)	35	0.3	
American copperhead (*Agkistrodon contortrix*)	52	10.5	
Bushmaster (*Lachesis muta*)	280–450	5.93	
Coral snake (*Micrurus fulvius*)	2–6	0.97	1.5 (mouse)
Tiger snake (*Notechis scutatus*)	70	0.04	0.045 (rabbit)
Indian cobra (*Naja naja*)	170–325	0.4	
Krait (*Bungarus caeruleus*)	10	0.09	0.04 (rabbit)
Eastern green mamba (*Dendroaspis angusticeps*)	80	0.45	
Beaked sea snake (*Enhydrina schistosa*)	7–20	0.01	

Table 1.1.23: Poisonous Effect of Selected Insecticides on Vertebrates
When the insecticide is orally administered, 50 % of the animals examined die (LD_{50}).
Figures in milligrams poison per kilogram body weight. The animal species examined are
in parentheses. For a comparison, see also Tables 1.1.21 and 1.1.22.

Figures from ALTMANN and DITTMER 1973

Insecticide	LD_{50} (mg/kg)	(Animal species)
Aldicarb	1	(Rat)
Aldrin	44	(Mouse)
	33	(Guinea pig)
Chlordane	430	(Mouse)
DDT	118	(Rat)
	400	(Guinea pig)
	150–400	(Mouse)
DDVP (Dichlorvos)	56– 80	(Rat)
Diazinon	76	(Rat)
Dieldrin	38	(Mouse)
	49	(Guinea pig)
	38– 87	(Rat)
Endrin	16– 36	(Guinea pig)
Heptachlor	68	(Mouse)
	116	(Guinea pig)
Lindane	86	(Mouse)
	100–127	(Guinea pig)
Mevinphos	3.7– 6.1	(Rat)
Parathion (E 605)	3.6	(Rat)
Perthane	9.34	(Rat)
Toxaphene	112	(Mouse)
	69	(Guinea pig)
	69	(Rat)
Zectran (mexacarbate)	15– 63	(Rat)

1.2 Movement, Locomotion

Table 1.2.0: Numbers that Amaze

This table brings together several interesting figures from Sect. 1.2, which are cited again in the individual tables below, where they are accompanied by references to the literature.

Number of wingbeats per second in insects		
	Chironomidae (*Forcipomyia*)	1,046/s
	Housefly	up to 330/s
Number of wingbeats in birds		
	Amethyst hummingbird	78/s
	House sparrow	13/s
	Pelican	1.3/s
Wing surface areas in birds		
	Griffon vulture	10,450 cm^2
	Ruby-throated hummingbird	10 cm^2
Greatest surface load in birds		
	Mute swan	167 N/m^2
	Albatross	155 N/m^2
Maximum speed, swimming		
	Swordfish	90 km/h
	Marlin	80 km/h
	Orca	65 km/h
Maximum speed, running		
	Cheetah	120 km/h
	Greyhound	110 km/h
	Red kangaroo	80 km/h
Maximum speed, flying		
	Common swift	180 km/h
	Spine-tailed swift	335 km/h
	Peregrine falcon, in a swoop	290 km/h
Maximum speed, during bird migration		
	Common swift	150 km/h
Annual migratory flight distance		
	Arctic tern	40,000 km
	Stork	20,000 km
Maximum jumping distance in animals		
	Red kangaroo	13.5 m
		= 9 times its body length
	Flea	60 cm
		= 200 times its body length

Table 1.2.1: Diameter of Muscle Fibers

Data for selected muscle types and organisms.

Figures from DAVID 1977

Species	Muscle	Diameter (μm)
Goldfish	Red fibers	36.0
	White fibers	49.4
African clawed frog	Leg muscle	169.0
Rock dove	Breast muscle	26.8
Shrew	Diaphragm	18.0
Mouse	Calf muscle	60.8
Rat	M. extensor digit. long.	85.0
	Diaphragm	34.0
Guinea pig	Diaphragm	25.0
Cat	Diaphragm	30.0
Pig	Diaphragm	60.0
Human	Diaphragm	34.0
	Intercostal muscles	50.4
	M. deltoideus	54.2

Table 1.2.2: Twitch Times for Muscles

The twitching of a muscle consists of contraction and relaxation. For contraction times, see Table 1.2.3. Listed in descending order from shortest to longest twitch.
On the physiology of muscles, see also SIEWING 1980.

After BERTELSMANN 1979, HEIDERMANNS 1957

Species	Muscle	Twitch time (s)
Housefly	Wing muscle	0.003
Bumblebee	Wing muscle	0.004
Bee	Wing muscle	0.005
Dragonfly	Wing muscle	0.039
Rock dove	Wing muscle	0.05
Grasshopper	Jumping muscle	0.1
Frog	Calf muscle	0.12
Bat	Wing muscle	0.5
European pond turtle	Biceps	0.9
Human	M. ciliaris	2– 4
Frog	Stomach	100–200
Pond mussel	Adductor muscle	200–600

Table 1.2.3: Contraction Times of Selected Muscles

Twitch times of muscles, which include both the contraction time and the relaxation time that follows it, are found in Table 1.2.2.

Figures from PROSSER 1973

Species	Muscle	Contraction time (ms)
Moon jelly		500–1,000
Edible snail	Tentacle retractor	2,500
Earthworm	Longitudinal muscle	70–80
Cockroach	Coxal muscle	4–10
Lobster	Fast remotor muscle	10
Lobster	Slow remotor muscle	50–60
Frog	Twitch fiber	22
Frog	Tonic fiber	46
Frog	Sartorius muscle	75
Toad	Semitendinosus muscle	800
Snake	Costocutaneous muscle	37–54
Turtle	Intestinal muscle	30,000
Finch	Wing muscle	14
Hummingbird	Wing muscle	8
Sloth	Gastrocnemius muscle	174
Guinea pig	Soleus muscle	82
Rat	Soleus muscle	36
Cat	Soleus muscle	70
Cat	Gastrocnemius muscle	40

Table 1.2.4: Maximum Muscular Strength of Selected Muscles

Force in N/cm² of a muscle cross-section. Compare this also to the so-called load-bearing performance in Table 1.2.5, a figure that combines maximum strength with load-bearing time.

Data after PROSSER 1973

Animal species	Muscle	Force (N/cm²)
Cockroach	Coxal muscle	7.8
Oyster	Adductor muscle	117.7
Lobster	Slow remotor muscle	27.5
Lobster	Fast remotor muscle	0.25
Frog	Sartorius muscle	19.6
Rabbit	Uterus	1.3
Dog	Smooth muscle trachea	7.8

(Cont.)→

Animal species	Muscle	Force (N/cm²)
Sloth	Gastrocnemius muscle	15.9
Sloth	Diaphragm	20.6
Rat	M. extensor digitorum	29.4
Rat	Gastrocnemius muscle	17.7
Rat	Soleus muscle	2.8
Rat	Cardiac muscle	0.2
Cat	Tenuissimus muscle	13.7
Cat	Duodenal sphincter	4.1

Table 1.2.5: Load-Bearing Record of Selected Muscles

The load-bearing performance describes the load in grams that can be borne by 1 cm² cross-section of a muscle for 1 h.

Data after HEIDERMANNS 1957

Species	Mucle type	Load-bearing performance
Frog	Calf muscle	4– 6
Toad	Calf muscle	44– 67
Human	Biceps	97–575
Pond mussel	Adductor muscle	1– 2 million

Table 1.2.6: Number of Wingbeats per Second in Insects

For the resulting speeds, see Table 1.2.10. For figures for birds, see Table 1.2.8. For greater detail on insect flight, see SIEWING 1980.

Data after BAUER et al. 1974, BERTELSMANN 1979, FRANZ 1959, HANKE et at. 1977, LAMPEITL 1982, MEYER 1964

Species	Frequency
Bee	
Worker bee	240–250
Soldier bee	285
Queen bee	226–253
Drone	207
Blowfly (*Calliphora*)	155
Bumblebee	130–250
Caddis fly	38
Cockchafer (May bug)	46

Species	Frequency
Common blue (butterfly)	11
Crane fly	44–73
Dragonfly (*Aeschna*)	22–28
Dragonfly (Broad-bodied chaser)	20
Horsefly	96
Housefly	180–330
Hummingbird hawk-moth	79–85
Ladybug	75–91
Mayfly (*Cloeon*)	41–44
Megaloptera (*Eristalis*)	177–188
Midge (*Forcipomyia*)	1,046
Migratory locust	20
Mosquito	278–307
Silkworm moth	8
Small white (butterfly)	9–12
Soldier beetle	69–87
Wasp	110

Table 1.2.7: Surface Areas and Surface Loads of Insect Wings

For data on birds, see Table 1.2.9. Listed according to increasing surface load.

After Oppenheimer and Pincussen 1925

Species	Weight (g)	Surface (cm²)	Surface load (cN/cm²)
Green lacewing	0.008	1.66	0.0047
Small white (butterfly)	0.100	15.80	0.0062
Small tortoiseshell	0.114	8.80	0.0111
Swallowtail butterfly	0.235	16.90	0.0136
Damselfly	0.045	3.05	0.0147
Mayfly (*Ephemera*)	0.0308	1.59	0.0190
Midge (*Chironomus*)	0.012	0.05	0.0235
Hummingbird hawk-moth	0.16	5.50	0.0284
Ladybug	0.031	0.93	0.0327
Dragonfly (*L. depressa*)	0.52	14.90	0.0342
Housefly	0.0115	0.31	0.0364
Wasp, worker	0.0780	1.36	0.0562
Privet hawk-moth	1.740	25.50	0.0669
Honeybee, worker	0.067	0.90	0.0730
Cockchafer (May bug)	0.667	8.15	0.0802

(Cont.) →

Species	Weight (g)	Surface (cm²)	Surface load (cN/cm²)
Bumblebee	0.118	1.18	0.0981
Earth-boring dung beetle	0.9039	5.90	0.1534
Water scavenger beetle	4.82	19.50	0.1603

Table 1.2.8: Wingbeat Frequencies in Birds

The speeds reached with these frequencies are found in Table 1.2.10. For figures on insects, see Table 1.2.6.

Data from DORST 1972, HERTEL 1963, HERZOG 1968, ZISWILER 1976

Species	Beats per second
Amethyst hummingbird	78 (!)
Blackbird	5.6
Carrion crow	5
Common buzzard	3
Common swift	12
Eurasian coot	5.8
European starling	5.1
Gray heron	2
Herring gull	3
House sparrow	13
Hummingbirds	30–50
Magpie	3
Mallard	5–10
Mute swan	2.7
Owls	4
Pelican	1.2
Peregrine falcon	4.3
Pheasant	9
Rock dove	8
Stork	2
Turkey	3

Table 1.2.9: Surface Areas and Surface Loads of Bird Wings

For corresponding figures for insects, see Table 1.2.7. For the wingspans of birds, see Table 1.1.9. Listed in descending order according to surface area.
As a point of comparison: gliders have a surface load of 25 N/m².

After DORST 1972, HERZOG 1968, OPPENHEIMER and PINCUSSEN 1925

Species	Surface area (cm²)	Surface load (N/m²)
Griffon vulture	10,450	71
Albatross	6,220	155
Eagle	5,930	69
Mute swan	5,760	167
Great bustard	5,550	74
Stork	5,520	66
Egret	3,590	38
Greylag goose	3,070	112
Great black-backed gull	2,060	77
Crow	2,050	30
Goshawk	2,050	40
Buzzard/Hawk	2,030	41
Herring gull	2,010	41
Peregrine falcon	1,540	37
Barn owl	1,160	24
Mallard	1,150	93
Rock dove	840	24
Northern lapwing	660	30
Magpie	640	32
Common kestrel	490	44
Blackbird	310	28
Great spotted woodpecker	240	30
European starling	190	43
Barn swallow	140	14
Common swift	130	29
Sparrow	120	24
Great tit	100	21
Chaffinch	100	21
Crest or Kinglet	50	11
Ruby-throated hummingbird	10	24

Table 1.2.10: Maximum Speeds of Selected Animals

All figures are in kilometers per hour. Some of the speeds given are only achieved for short stretches. The great differences in the data given in the literature on bird flying speeds are based on the fact that no mention is made whether the cited speeds were achieved with or without tailwinds.

Data after Bauer et al. 1974, Bertelsmann 1979, Franz 1959, Grzimek 1970, Hanke et al. 1977, Heinroth 1955, Hertel 1963, Herzog 1968, Hesse and Doflein 1935, Kästner 1973, Krumbiegel 1953, Lampeitl 1982, Mann 1979, Mossmann and Sarjeant 1983, Niethammer 1979, Schwerdtfeger 1975, Slijper 1967, Wood 1982, Wunderlich and Gloede 1977, Ziswiler 1976

Species	km/h	Species	km/h
Swimming		Coyote	60
Common eider	35	Deer	67
Dolphin	46	Dromedary	50
Eel	12	Eastern green mamba	11.3
Fin whale	55	Edible snail (*Helix pomatia*)	0.00324
Gentoo penguin	36	Gazelle	75
(under water)		Giant tortoise	0.33
Giant squid	40	Giraffe	51
Grass snake	2.9	Gnu	60
Gray whale	7.5	Greyhound	110
Great diving beetle	1.8	Hare	65
Human (100 m freestyle)	7	Herbivorous dinosaurs	6
Marlin	80	Hippopotamus	48
Orca	65	Horse (gallop)	36
Polar bear	10	Horse (trot)	13.5
Ringed seal	32	House cat	48
Salmon	39	Human (100 m race)	36
Sea snakes	3.6	Human (flying start)	44.6
Shark	36	Lion	75
Swordfish	90 (!)	Mole	4
Trout	35	Ostrich	72
Tuna	75	Pill millipede (*Glomeris*)	0.013
		Pine moth larva	0.018
Crawling, running		Planarian	0.009
African elephant	39	Polar bear	65
Antilopes	70	Racehorse	69
Bison	55	Rat	9.6
Black mamba	17.6	Rattlesnake	4
Camel	15	Red deer (Wapiti)	70
Centipede (*Lithobius*)	1	Red kangaroo	80
Cheetah	120 (!)	Rhinoceros	51
Common house mouse	12	Roadrunner	24
Common vole	8	Shore crab	3.6

Species	km/h	Species	km/h
Sled dogs	23.4	Goshawk	42
Sloth	0.146	Gray partridge	85
Spider (*Tegenaria atrica*)	1.3	Great cormorant	70
Starfish	0.000576	Green lacewing	2.2
Theropods (dinosaurs)	16.5	Hawk moth	54
Western green lizard	29	Homing pigeon	80
Wild ass	65	Homing pigeon	177
Wild horse	60	(with a tailwind)	
Wolf	60	Hornets	22
Zebra	65	Horsefly	50
		House sparrow	45
Flying		Housefly	8.2
Albatross	70	Hummingbird hawk-moth	18
Bat	50	Mallard	104
Bee	29	Mergus	129
Blowfly	11	Migratory locust	16
Bumblebee	18	Mosquito	1.4
Cockchafer (May bug)	11	Peregrine falcon, in a swoop	290
Common buzzard	45	Small passerines	up to 61
Common swift	180	Small white (butterfly)	14
Crows	59	Sparrowhawk	58
Dragonfly	30	Spine-tailed swift	335 (!)
European kingfisher	58	Stork	45
European starling	81	Swallow	65
Falcons	79	Swan	50
Flyingfish	75	*Tabanus bovinus* (horsefly)	22
Geese	91	Tiger beetle	8

Table 1.2.11: Jumping Distances of Selected Animals

Listed in ascending order according to jump in relation to body length (BL).

Data after COCHRAN 1970, GRZIMEK 1970, HESSE and DOFLEIN 1935, KRUMBIEGEL 1953

Species	Distance jumped (m)	Relation to body length (BL)
Lion	4–5	2–3 times BL
Tiger	5	2–3 times BL
Fox	2.8	4.3 times BL
Red deer (Wapiti)	11	4.5 times BL
Roadrunner	3	6 times BL
Impala antelope	10	6 times BL
Least weasel	1.2	6 times BL
Kangaroo	6–10 max. 13.5	7 times BL
Wood mouse	0.7	8 times BL
Bullfrog	2	10 times BL
Gibbon	12	13 times BL
Jerboa	2.5	15 times BL
Grasshopper	2	30 times BL
Agile frog	2	33 times BL
Northern cricket frog (*Acris crepitans*)	1.25	36 times BL
Sharp-nosed ridged frog (*Rana [Ptychadena] oxyrhynchus*)	3.6	45 times BL
Flea	0.6	200 times BL
In comparison: Human, current world record	8.9	5 times BL

Table 1.2.12: Flight Speeds During Bird Migration

For maximum speeds, see Table 1.2.10. For annual flight distances, see Table 1.2.13. Listed here in ascending order, according to speed.

After CREUTZ 1965, DORST 1972, SCHÜZ 1971

Species	Km/h
Red-backed shrike	40.0
Sparrowhawk	41.4
Barn swallow	44.0
Honey buzzard	45.0
Blue-headed wagtail	46.0–48.0
Wood pigeon, crane, great black-backed gull, hooded crow	50.0
Eurasian siskin	51.0

Species	km/h
Rook	52.2
Chaffinch	52.5
Meadow pipit	53.0
Peregrine falcon	59.2
Crossbill	59.8
Turtle dove	61.0–72.0
Jackdaw	61.6
European starling	74.2
Plover family (Charadriidae)	90.0
Common teal	120.0
Common swift	150.0

Table 1.2.13: Flight Distances During Bird Migration

The figures represent an outbound or return flight, which means that the annual distance flown is twice as far. For flight speed during bird migration, see Table 1.2.12. Listed according to distance covered.

Data after BERNDT and MEISE 1959, CREUTZ 1956, SCHÜZ 1971, ZISWILER 1976

Species	Distance migrated (km)
European starling	1,400
Song thrush	2,600
Northern lapwing	3,000
Common quail	5,000
Greenland wheatear	5,000
Japanese snipe	5,000
Lesser whitethroat	6,000
Crane	6,500
Cuckoo	9,500
Barn swallow	10,000
Stork	10,000
Amur red-footed falcon	10,000
Eurasian dotterel	12,000
American golden plover	15,000
Short-tailed shearwater	17,500
Pectoral sandpiper	20,000
Arctic tern	20,000

Table 1.2.14: Migratory Times of Selected Birds Native to Germany

Roman numerals are used to indicate the months. The Arabic numbers 1, 2, and 3 after the slash stand for the first, second, and third "third" of the month.

Figures from FLINDT 1985, SCHÜZ 1971

Species	Spring migration (months)	Autumn migration (months)
Barn swallow	IV/3	IX–X
Black redstart	III/1	X
Blackcap	IV/2–V	IX–X
Chiffchaff	III/2	IX–XI
Common quail	V/3	VIII/2–X
Common redstart	IV/1	X
Common swift	IV/3–V/1	VIII/1
Crane	III/3–IV	XI
Cuckoo	IV	VIII–IX
Eurasian curlew	IV	VIII/3–X
Fieldfare	III–V/1	X–XI
Garden warbler	V	IX
Garganey	III/3–IV	VIII–IX
Golden oriole	IV/3–V/1	IX/1
Gray heron	III–IV	IX–X
Great reed warbler	IV/3	VIII–IX
Greylag goose	III	IX–X
Hoopoe	IV/2	IX–X
Lesser whitethroat	IV/2	IX–X
Nightingale	IV/3	VIII/3–IX
Northern lapwing	III/1	VIII–X
Sedge warbler	IV	IX–X
Snipe	III–IV	IX–XI
Song thrush	II–III	IX–X
Stork	II–IV	VII–VIII
Tree pipit	IV/2, 3	IX
White wagtail	III–V	X
Whitethroat	IV/3–V	IX
Willow warbler	IV/2	IX–X
Wood warbler	IV/3–V/1	VIII/3–IX
Woodlark	II–III	IX–X
Wryneck	IV/2–V/1	VIII/3–IX

1.3 Alimentation, Digestion

Table 1.3.0: Numbers that Amaze

This table brings together interesting figures from the Sects. 1.3 Alimentation, Digestion, 1.4 Metabolism, and 1.5 Excretion. These numbers are cited again in the individual tables, where they are accompanied by references to the literature. *BL* body length, *BW* body weight.

Daily dietary needs as share of body weight (BW)	Etruscan shrew	200% of BW
	Hummingbird	200% of BW
	Asian elephant	1% of BW
Length of intestine		
	Sperm whale (absolute)	288 m
	– As compared to BL	24 times
	Minke whale as compared to BL	56 times
Capacity of a cow's stomach		252.5 l
Capacity of a horse's appendix		33.5 l
Capacity of a horse's large intestine		96 l
Body temperature	Shrew	42 °C
	Blackbird	43.6 °C
	Robin	44.6 °C
	Hedgehog during hibernation	6 °C
Length of hibernation	Hazel dormouse, fat dormouse	6–7 months
Number of renal glomeruli	Cattle	8,000,000
Daily urine production	Cattle	up to 20 l
	Horse	up to 10 l
Relative weight of liver (as % of BL)		
	Kitefin shark (*Scymnorhinus*)	22.7%

Table 1.3.1: Absorption of Nutrients in the Body

RQ respiratory quotient
For more detail on digestion, see SIEWING 1980.

After PENZLIN 1970

	O_2 consumption in cm^3 per g nutrient	CO_2 released in cm^3 per g nutrient	RQ CO_2/O_2	Energy in kcal per g (kJ/g) (kJ/l O_2)	Caloric equivalent kcal/l O_2
Carbohydrates 820	820	1	4.1 (17.2)	5.05 (21.1)
Fats 2,020	1,430	0.71	9.4 (39.3)	4.65 (19.5)
Proteins 960	770	0.8	4.3 (18.0)	4.49 (18.8)
(physiological oxidation)				

Table 1.3.2: Required Daily Food Intake

Figures given are percentage of body weight (BW). Listed according to increasing food requirements. Food requirements are related to the basal metabolic rate. See Table 1.4.1.

After HESSE and DOFLEIN 1935, NIETHAMMER 1979, SCHWERDTFEGER 1963, SLIJPER 1967

Species	Food requirements (as % of BW)
Anaconda	0.013
Asian elephant	1.0
Bear	2.0
Tiger	2.8
Lion	2.9
Cattle	3.0
Chicken	3.5
Buzzard/Hawk	4.5
Little owl	6.5
Common kestrel	8.0
Song thrush	10.0
European starling	11.9
Crest or Kinglet	18.0
Least weasel	25.0
Blue tit	30.0
Mouse	40.0
Mole	100.0
Shrews	100.0
Hummingbird	200.0
Pygmy mouse	200.0

Table 1.3.3: Dental Formulae of Selected Mammals

The numbers above the fraction bar represent the number of the following distinct types of teeth in the upper jaw: incisors, canines, premolars, and molars, respectively. The numbers below the fraction bar designate the same types of teeth in the lower jaw.

Data after BRINK 1975, BROHMER 1977, GIERSBERG and RIETSCHEL 1968, GRZIMEK 1970, KRUMBIEGEL 1953

Species	Dental formula
Apes	$\frac{2\ 1\ 2\ 3}{2\ 1\ 2\ 3}$
Arvicolinae	$\frac{1\ 0\ 0\ 3}{1\ 0\ 0\ 3}$
Bats:	
Horseshoe bats	$\frac{1\ 1\ 2\ 3}{2\ 1\ 3\ 3}$
Mouse-eared bats	$\frac{2\ 1\ 3\ 3}{3\ 1\ 3\ 3}$
Bears	$\frac{3\ 1\ 4\ 2}{3\ 1\ 4\ 3}$
Beaver	$\frac{1\ 0\ 1\ 3}{1\ 0\ 1\ 3}$
Beech marten	$\frac{3\ 1\ 4\ 1}{3\ 1\ 4\ 2}$
Camel	$\frac{1\ 1\ 3\ 3}{3\ 1\ 2\ 3}$
Cat	$\frac{3\ 1\ 3\ 1}{3\ 1\ 2\ 1}$
Cattle	$\frac{0\ 0\ 3\ 3}{3\ 1\ 3\ 3}$
Chamois	Same as cattle
Common or Harbor seal	$\frac{3\ 1\ 4\ 1}{2\ 1\ 4\ 1}$
Deer	Same as cattle
Dog	Same as bears
Dormice	Same as beaver
Elephant	$\frac{1\ 0\ 3\ 3}{0\ 0\ 3\ 3}$

(Cont.) →

Species		Dental formula			
Ermine		$\underline{3}$	$\underline{1}$	$\underline{3}$	$\underline{1}$
		3	1	3	2
Eurasian badger	Same as pine marten				
Fat dormouse	Same as beaver				
Fox	Same as bears				
Giraffe	Same as cattle				
Goat	Same as cattle				
Guinea pig	Same as beaver				
Hamster	Same as arvicolinae				
Hare		$\underline{2}$	$\underline{0}$	$\underline{3}$	$\underline{3}$
		1	0	2	3
Hazel dormouse	Same as beaver				
Hedgehog		$\underline{3}$	$\underline{1}$	$\underline{3}$	$\underline{3}$
		2	1	2	3
Hippopotamus		$\underline{2}$	$\underline{1}$	$\underline{4}$	$\underline{3}$
		2	1	4	3
Horses		$\underline{3}$	$\underline{1}$	$\underline{3}$	$\underline{3}$
		3	1	3	3
Human	Same as apes				
Hyena		$\underline{3}$	$\underline{1}$	$\underline{3}$	$\underline{1}$
		3	1	3	1
Ibex	Same as cattle				
Least weasel	Same as ermine				
Lion	Same as cat				
Llama		$\underline{1}$	$\underline{1}$	$\underline{2}$	$\underline{3}$
		3	1	2	3
Lynx		$\underline{3}$	$\underline{1}$	$\underline{2}$	$\underline{1}$
		3	1	2	1
Marmot	Same as squirrel				
Mole		$\underline{3}$	$\underline{1}$	$\underline{4}$	$\underline{3}$
		3	1	4	3

Species	Dental formula			
Murinae	Same as arvicolinae			
New World monkeys	Same as prosimians			
Old World monkeys	Same as apes			
Old World Porcupine	Same as beaver			
Opossum	5	1	3	4
	4	1	3	4
Pigs	Same as mole			
Pine marten	Same as beech marten			
Polecat	Same as ermine			
Prosimians	2	1	3	3
	2	1	3	3
Rabbit	Same as hare			
Raccoon	3	1	4	2
	3	1	4	2
Rats	Same as arvicolinae			
Rhinoceros	0	0	4	3
	0	0	3	3
Roe deer	Same as cattle			
Sheep	Same as cattle			
Shrews				
Genus *Sorex*	3	1	3	3
	1	1	1	3
Genus *Neomys*	3	1	2	3
	1	1	1	3
Genus *Crocidura*	3	1	1	3
	1	1	1	3
Squirrel	1	0	2	3
	1	0	1	3
Tapir	3	1	4	3
	3	1	3	3
Tiger	Same as cat			
Walrus	3	1	3	2
	3	1	3	1
Wild cat	Same as cat			
Wolf	Same as bears			

Table 1.3.4: Intestinal Length of Selected Vertebrates

For relative intestinal length (in relation to body length), see Table 1.3.5.

Data after BERTELSMANN 1979, BUDDENBROCK 1956, HALTENORTH 1977, KRUMBIEGEL 1953, MEYER 1964, NIETHAMMER 1979

Species	Length (m)
African elephant	19.0
Bat	0.25
Blue whale	120.0
Brown bear	17.0
Camel	42.0
Cat	2.1
Cattle	51.0
Chicken	1.8
Dog	5.0
Eurasian water shrew	0.26
Goat	32.0
Hedgehog	1.7
Horse	30.0
Human	6.0–8.0
Lion	6.9
Mole	1.2
Mouse	0.5
Pig	22.0
Rabbit	5.6
Rat	1.3
Roe deer	12.0
Sheep	31.0
Siren ("sea cow")	44.0
Sperm whale	288.0 (!)
Turtle	0.9
Wolf	5.8

Table 1.3.5: Relative Intestinal Length of Selected Animals

Numbers indicate relation of intestinal length (*IL*) to body length (*BL*) (how many times longer the intestine is than the body). For absolute lengths, see Table 1.3.4. Among the herbivores (long intestines) and carnivores (short intestines), respectively, the intestines of smaller animals are relatively longer than those of their larger counterparts. The sperm whale represents an exception to this pattern among whales, probably due to its feeding on octopus and squid.

Data after BAUER et al. 1974, BUDDENBROCK 1956, BERTELSMANN 1979, HALTENORTH 1977, HESSE and DOFLEIN 1935, KRUMBIEGEL 1953, *Lexikon der Biologie*, 1964, NIETHAMMER 1979, ZISWILER 1976

Species	Relation of IL to BL (times)	Species	Relation of IL to BL (times)
Bat	1.9	Leguan	0.3
Blue whale	4.5	Lion	3.9
Brown bear	8	Minke whale	56
Camel	12	Mole	8–11
Cat	3–4	Mouse	8
Cattle	22–29	Mudpuppy	3
Chicken	1.8	Nile crocodile	1.8
Cockchafer (May bug)	7	Orangutan	4.5
Common or Harbor seal	28	Pig	14–15
Common shrew	1.3	Pike	1
Dog	5–6	Porpoise	20
Dolphin	15	Python	1.8
Dung beetle (*Scarabaeus*)	13	Rabbit	10
Elephant	7	Rat	9–11
Eurasian water shrew	3	Rock dove	7
European pond turtle	9	Roe deer	11
Fox	2.9	Sea lion	37
Great diving beetle	1	Sheep	27
Hedgehog	6.6	Snakes	1
Hippopotamus	12–14	Sperm whale	16–24
Horse	12	Tiger	5
Human	7.5	Turtles	5
Ibex	27	Wolf	4.7

Table 1.3.6: Proportion of the Total Volume of the Intestinal Tract Made Up by Stomach, Small Intestine, and Appendix in Selected Mammals

Figures represent percentages (%) of total volume. For absolute numbers, see also Table 1.3.7.

From BUDDENBROCK 1956, modified

Species	Stomach (%)	Small intestine (%)	Appendix (%)
Cattle	70.8	18.5	2.8
Dog	62.3	23.3	1.3
Goat	66.9	20.4	2.3
Guinea pig	19.4	31.1	21.4
Horse	8.5	30.2	15.9
Muskrat	27.2	16.5	37.8
Nutria	19.4	16.7	44.4
Pig	29.2	33.5	5.6
Rabbit	15.1	37.7	31.4
Sheep	66.9	20.4	2.3
Squirrel	42.1	33.4	14.0
White rat	24.2	43.7	16.8

Table 1.3.7: Capacity of the Digestive Tract and its Parts in Selected Domestic Animals

From KOLB 1974

Species	Absolute capacity (l)	Relative capacity (%)
Cat	0.579	
Stomach	0.341	69.5
Small intestine	0.114	14.6
Large intestine	0.124	15.9
Dog	6.95	
Stomach	4.33	62.3
Small intestine	1.62	23.3
Appendix	0.09	1.3
Large intestine and rectum	0.91	13.1
Pig	27.45	
Stomach	8.00	29.2
Small intestine	9.20	33.5
Appendix	1.55	5.6
Large intestine and rectum	8.70	31.7

Species	Absolute capacity (l)	Relative capacity (%)
Horse	211.34	
Stomach	17.96	8.5
Small intestine	63.82	30.2
Appendix	33.54	15.9
Large intestine	96.02	45.4
Sheep	44.20	
Stomach	29.60	66.9
Small intestine	9.00	20.4
Appendix	1.00	3.3
Large intestine and rectum	4.60	10.4
Cattle	356.40	
Stomach	252.50	70.8
Small intestine	66.00	18.5
Appendix	9.90	2.8
Large intestine and rectum	28.00	7.9

Table 1.3.8: pH Optimum of Selected Digestive Enzymes in Mammals

Data after BAUER et al. 1974, BUDDENBROCK 1956, *Documenta Geigy* 1975, GANONG 1972, KARLSON 1974

Enzyme	pH value
Aminopeptidase	8.0
Carbohydrases	7.0
Cathepsin	4.0–6.0
Chymotrypsin	7.5–8.5
Enteropeptidase	6.0
Gastricin	3.0–3.5
Lipases from carnivores	8.0
Lipases from herbivores	8.4–9.0
Maltase	6.6–7.0
Pancreatic α-amylase	6.9–7.1
Pepsin A	1.5–2.5
Ptyalin	6.7
Trypsin	7.5–8.5

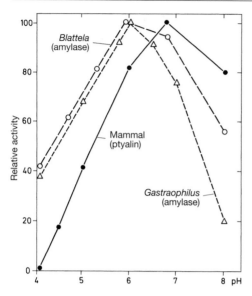

Fig. 1.3.8: Dependency of activity of selected amylases on pH value. See also Table 1.3.8. From PENZLIN 1980, p. 33.

Table 1.3.9: pH Value of the Saliva of Selected Mammals

From BUDDENBROCK 1956, modified

Species	pH value
Cattle	7.99–8.27
Dog	7.34–7.80
Horse	7.31–7.80
Pig	7.15–7.47
Human	5.93–6.86

Table 1.3.10: pH Values in the Intestinal Tract of Selected Birds

For figures on mammals, see Table 1.3.11.

Figures from BUDDENBROCK 1956, KOLB 1974

	Crop	Gizzard	Duodenum	Ileum	Large intestine	Appendix
Chicken	4.51	2.6	5.76–6.01	5.78–6.42	6.26	5.71
Rock dove	4.28	2.0	5.23–5.39	5.32–5.89	5.43	
Pheasant	5.78	2.06	6.01–6.19	6.18–6.81	6.61	5.39
Duck	4.92	2.33	6.01–6.19	6.71–6.95	6.37	5.88

Table 1.3.11: pH Values in the Intestinal Tract of Selected Mammals

Figures from various authors in BUDDENBROCK 1956, modified

Species	Stomach	Duode-num	Jeju-num	Ileum	Cecum	Colon	Rectum
Horse	4.48	7.5	7.79		7.26	6.95	
Horse	4.45	7.13	7.46		7.23	7.09	6.24
Pig	5.46	6.63	7.44		7.24	6.85	
Rat	6.98		7.32	7.28	7.15		
Rat	3.2–4.6	6.5– 7.2			5.5–7.3	5.6–7.1	
Cattle	4.95[a]	5.98–7.49		7.54	8.68	7.15	
Cattle	5.53	6.83	8.55	7.93	9.41		
Suckling calf	4.46	6.8–7.1	6.9–7.2	7–7.3	7.2–7.5	7.1–7.5	7.1–7.5

[a]Abomasum

Table 1.3.12: pH Values in the Mid-intestine of Selected Invertebrates

From BUDDENBROCK 1956, modified

Species	pH value
Bedbug	6.2
Bee	6.3
Butterfly (*Pyrameis*)	8.4
Cockroach (*Blatella*)	6.2
Cricket	7.6
European rhinoceros beetle	10.0–10.4
Grasshopper (*Tettigonia*)	5.9
Greenbottle blowfly (*Lucilia*)	7.8–8.0
Louse	7.2
Meal moth	8.4
Paper wasp (*Polistes*)	7.3
Primitive insects (*Ctenolepisma*)	4.8–7.0
Soft-shell clam (*Mya arenaria)*	6.7
Walking stick	6.3

Table 1.3.13: Duration of Intestinal Passage in Selected Vertebrates

This table shows the speed of passage of food through the intestine and the start and end of defecation.

Data after BUDDENBROCK 1956, HALTENORTH 1977

Species	Speed of passage (cm/h)	Start of defecation		End of defecation	
Bat	3.3	41	min	7	h
Camel	9.7			18	days
Cattle	17.7	25	h	12	days
Dog	7.7	17	h	2.7	days
Horse	35.7	24	h	3.5	days
Mouse	6.4	3	h	8.5	h
Pig	22.9	13	h	4	days
Rabbit	4.7	5	h	5	days
Rat	1.1	6.5	h	5	days
Sheep	8.1	12	h	16	days

Table 1.3.14: Relative Weight of Liver in Selected Vertebrates

As percentage of body weight (BW). Figures given are mean values.

After BUDDENBROCK 1956, HALTENORTH 1977, KOLB 1974

Species	Liver (as % of BW)	Species	Liver (as % of BW)
African elephant	1.6	Human	3.0
Cat	3.6	Kitefin shark	22.7
Cattle	1.2	(*Scymnorhinus*)	
Chicken	2.7	Lion	3.2
Chimpanzee	2.8	Mouse	7.1
Common vole	4.6	Pig	2.4
Dog	2.9	Rabbit	2.7
Duck	2.4	Rat	3.8
Electric rays	5.7	Rock dove	3.0
Goat	1.3	Sheep	1.3
Golden or Syrian hamster	5.2	Sparrow	5.6
Gorilla	5.1	Squirrel	2.2
Guinea pig	3.9	Stingray	14.9
Hippopotamus	1.8	(*Trygon*)	
Horse	1.4	Toad	2.8
Houndshark	9.7	Wolf	2.8

Table 1.3.15: Daily Bile Production in Selected Vertebrates

Figures express daily bile production in relation to hepatic weight and body weight.

From BUDDENBROCK 1956, modified

Species	cm³ bile per g liver in 24 h	cm³ bile per kg body weight in 24 h
Cat	0.45	14.0
Cattle	0.86	15.4
Chicken	0.58	14.2
Dog	0.36	12.0
Duck	0.91	40.1
Goat	0.87	11.8
Guinea pig	6.50	228.0
Hamster	3.03	72.3
Horse	1.46	20.8
Human	0.64	16.0
Mouse	0.63	34.9
Pig	1.04	25.2
Rabbit	4.53	118.0
Rat	1.25	47.1
Rock dove	1.64	40.1
Sheep	0.98	12.1

1.4 Metabolism

Table 1.4.1: Basal Metabolic Rate of Selected Animals

Figures give oxygen consumption in cubic centimeters of O_2 per gram body weight and per hour. 1 cm³ O_2 is the equivalent of 1.43 mg. The basal metabolic rate in homoiothermic small animals is, relative to body weight, higher than in comparable large animals, since surface area only increases when squared, but weight (= volume) increases when cubed (see Fig. 1.4.1). Relative to body surface area, the basal metabolic rate is roughly the same. For greater detail, see also SIEWING 1980.

Date after FLOREY 1976, HEIDERMANNS 1957, PROSSER 1973, WHITTOW 1970

Species	O_2 consumption (cm³ O_2 per g BW and per h)	Species	O_2 consumption (cm³ O_2 per g BW and per h)
Invertebrates		Cattle	0.124
Amoeba	0.2	Chaffinch	3.3
Blowfly	1.7	Chicken	0.43
Blue mussel	0.005–0.076	Common house mouse	1.7
Butterfly (*Vanessa*)	0.6	Dog	0.36
Crayfish	0.1	Eel	0.088
Earthworm	0.065	Elephant	0.07–0.11
Freshwater snail	0.011	Frog (*Rana*)	0.055
Lobster	0.5	Frog Tadpole (*Rana*)	0.065
Lugworm	0.031	Goldfish	0.042–0.07
Mealworm	0.182	Grass snake	0.07
Octopus (*Octopus*)	0.09–0.28	Guinea pig	0.6–0.9
Paramecium	1.0–1.3	Hare	0.96
Pond mussel	0.002	Horse	0.03–0.04
Roundworm	0.5	Human	0.21–0.24
Sea anemone	0.013	Hummingbird	4.0
Sea urchin	0.004–0.027	Kangaroo	0.29
Shore crab	0.094	Lizard (*Lacerta*)	0.245
Sponge (*Suberitis*)	0.24	Pig	0.35
Starfish	0.03–0.4	Python	0.02
Woodlouse	0.096	Rabbit	0.55
		Rat	0.88–0.95
Vertebrates		Sheep	0.25
Alligator	0.079	Shrew	7.0–10.6
Bat	1.74	Sparrow	3.53
Camel	0.099	Trout	0.205–0.22
Carp	0.067		
Cat	0.45		

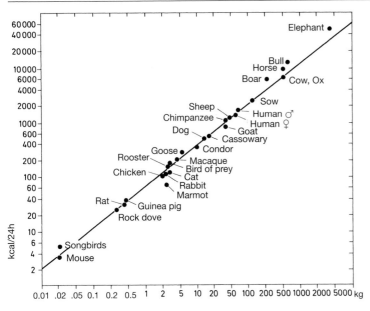

Fig. 1.4.1: Dependency of basal metabolic rates on the body weight of selected homoiothermic animals on a double logarithmic scale. See also Table 1.4.1. From Penzlin 1980, p. 49

Table 1.4.2: Body Temperatures of Selected Homoiothermic Animals

Temperatures given are mean values of rectal readings. For daily variations in body temperature, see Table 1.4.3.

Data after Bertelsmann 1979, Dorst 1972, Kolb 1974, Meyer 1964, Niethammer 1979, Penzlin 1977, Precht et al. 1955, Slijper 1967, Wittke 1972, Ziswiler 1976

Species	Body temperature (°C)	Species	Body temperature (°C)
Birds		European starling	41.3
Albatross	40.7	Goose	40.7
Blackbird	43.6	Goshawk	41.9
Black-headed gull	41.4–42.1	Great spotted woodpecker	42.1–43.3
Carrion crow	42.0	King penguin	37.7
Chicken	41.5	Ostrich	37.4
Common buzzard	42.0	Robin	44.6
Common swift	44.0	Rock dove	41.8
Duck	40.5–42.5	Rufous hummingbird	39.0

(Cont.) →

Species	Body temperature (°C)	Species	Body temperature (°C)
Sparrow	41.4	Hazel dormouse	30.0
Sparrowhawk	41.2	Hedgehog, awake	35.0
Stork	40.0	Hedgehog, in hibernation	6.0
Tawny owl	41.0	Hippopotamus	35.4
Winter wren	41.8	Horse	37.6–37.8
		Human	36.2–37.8
		Kangaroo	36.8
Mammals		Long-beaked echidna	30.0
Bat	31.0	Marmot	36.0
Capuchin monkey	38.5	Mole	39.4
Cat	38.8–39.0	Mouse	38.0
Cattle	38.5	Pig	39.0
Chimpanzee	37.0	Rabbit	39.2–39.6
Dog	38.3–39.0	Rat	38.1
Dromedary	38.1	Reindeer	37.5–39.8
Elephant	36.2	Sheep	39.5
Fox	38.5	Shrew	42.0
Goat	40.0	Spiny anteater	32.0
Guinea pig	36.0–39.2	Whale	36.5

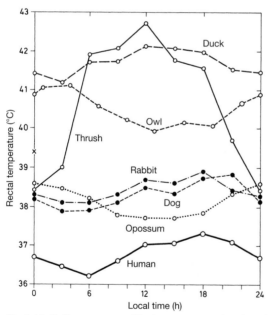

Fig. 1.4.2: Daily movement of body temperature in selected animals. Note the difference between nocturnal animals (owl, opossum) and their diurnal counterparts.
From PRECHT, CHRISTOPHERSEN and HENSEL 1955

Table 1.4.3: Size of Daily Variations in Body Temperature in Selected Homoiothermic Animals

For mean temperature values, see Table 1.4.2.

Data after NIETHAMMER 1979, PRECHT et al. 1955

Species	Daily variation (°C)
Camel	3.0
Dog	0.9
Duck	0.9
European starling	3.3
Goshawk	1.8
Guinea pig	0.8
Gull	1.7
Hedgehog	2.0
Human ♀	1.4
Human ♂	1.2
Long-beaked echidna	3.0
Owl	0.9
Rabbit	0.8
Rhesus monkey	3.6
Rock dove	1.4
Thrush	4.0

Table 1.4.4: Preferred Temperatures of Selected Animals

Preferred temperatures refer to those temperatures at which an animal most prefers to stay and which it actively seeks.

Data after BELLAIRS 1971, GRZIMEK 1970, HERTER 1962, KÄSTNER 1972, KEMPER and DÖHRING 1967, MEYER 1964

Species	Preferred temperature (°C)
Invertebrates	
Bedbug	33.0
Body louse	32.5
Crayfish (*Cambarus*)	25.0
Darkling beetle (*Adesmia metallica*)	49.0
Dog flea	38.0
Earwig	32.0
Field cricket	37.0
Great diving beetle	25.0
Ground beetle (*Carabus granulatus*)	30.0

(Cont.) →

Species	Preferred temperature (°C)
Katydid	40.7
Paramecium	23.0–25.0
Snow flea (*Boreus hiemalis*)	9.8
Tiger beetle	44.0
Wasp, queen	38.1
Wasp, worker	35.4
Winter moth	6.0
Vertebrates	
American alligator	32.0–35.0
Carp	21.3
Chameleon	45.0
Common house mouse	37.0
Common toad	27.9
Common vole	35.0
Crossed viper	33.0
Gecko (*Tarentola*)	38.0
Giant tortoise	28.0–33.0
Goldfish	18.6
Grass snake	34.3
Guinea pig	36.0
Guppy	24.6
Hermann's tortoise	25.0–33.0
Marine iguana	22.0–27.0
Rainbow trout	10.5
Sand lizard	38.8
Slowworm	28.4
Water vole	29.0

Table 1.4.5: Duration of Hibernation (Torpor, Winter Rest) of Selected Vertebrates Native to Germany

Data after BERTELSMANN 1977, FLINDT 1989, MERTENS 1947, MEYER 1964

Species	Duration (months)
Bats	5–6
Common frog	4–5
Common lizard	5–6
Common toad	4–5
Common wall lizard	5–6
Crossed viper	4–5
Fat dormouse	6–7
Grass snake	4–5
Hamster	2–3.5
Hazel dormouse	6–7
Hedgehog	3–4
Marmot	5–6
Sand lizard	5–6
Slowworm	4–5
Smooth newt	3–4
Squirrel	2–3.5
Tree frog	5–6

1.5 Excretion

Table 1.5.1: Number of Glomeruli in the Kidneys of Selected Vertebrates

The absolute number (n) is accompanied by a relative measurement, the number of glomeruli per gram of body weight (BW). Listed in ascending order, according to absolute number of glomeruli.

Values from HALTENORTH 1977, OPPENHEIMER and PINCUSSEN 1930, PORTMANN 1959, STURKIE 1976, ZISWILER 1976

Species	Glomeruli (n)	Glomeruli (per g BW)
Newt	400	80
Frog	2,000	70
Mouse	10,000	670
Bat	34,000	9,675
Rat	52,000	123
Wood mouse	72,000	3,092
Shrew	82,000	12,660
Squirrel	100,000	304
Mole	130,000	1,855
Hare	163,000	87
Hedgehog	186,000	238
Rabbit	285,000	188
Baboon	305,000	156
Cat	460,000	200
Dog	500,000	50
Rock dove	535,000	1,274
Fox	695,000	203
Chicken	840,000	336
Eurasian badger	962,000	80
Roe deer	1,500,000	60
Goose	1,659,000	307
Duck	1,989,000	542
Human	2,000,000	27
Dolphin	3,990,000	139
Deer	4,590,000	32
Common or Harbor seal	4,643,000	268
Cattle	8,000,000	12

Table 1.5.2: Urine Production of Selected Mammals

Figures given represent mean values. They are subject to a great degree of fluctuation, given changes in feed or intake of liquids. On the mechanism of urine production, see SIEWING 1980. For humans, see also Table 4.5.15.

From BUDDENBROCK 1956, KOLB 1974

Species	Amount in 24 h	Amount in 24 h (cm³ per kg BW)
Brown bear	2.2 l	21.4
Cat	100– 200 cm³	80.0
Cattle	12–20 l	24.0
Coyote	560 cm³	58.3
Dog (large)	0.5–2 l	40.0
Dog (small)	0.04–0.2 l	40.0
Fox	1.3 l	32.8
Horse	6–10 l	8.4
Human	1–1.6 l	14–23.0
Hyena	545 cm³	0.8
Pig	2–4 l	40.0
Sea lion	604 cm³	0.9
Sheep	1–1.5 l	10–15.0

Table 1.5.3: Maximum Urine Concentration in Selected Mammals

The level of urine concentration is in part dependent upon the length of Henle's loop in nephrons.

Data after PROSSER 1973

Species	Maximum concentration (mOsm)
Camel	2,500–3,170
Cat	2,100–3,250
Common or Harbor seal	2,150–2,420
Dog	2,000–2,660
European ground squirrel (suslik)	3,900
Golden or Syrian hamster	3,000
Harbor porpoise	1,830
Hopping mouse (*Notomys*)	5,540–6,400
Human	1,400–1,480
Kangaroo rat (*Dipodomys*)	5,500–6,000
Pig	1,100
Rabbit	1,500–1,910

(Cont.) →

Species	Maximum concentration (mOsm)
Rat	2,560–3,000
Rhesus monkey	1,190
Whale	900–1,340

Table 1.5.4: Composition of the Excrement of Selected Vertebrates

Figures represent the share of ammonia, urea, uric acid, creatine, and creatinine in excrement, given as a percentage of nitrogenous substances. On the ureic contents of urine, see Table 1.5.5

From BUDDENBROCK 1956, modified

	Ammonia	Urea	Uric acid	Creatine + Creatinine
Shark	2 – 11	81 – 89	0.3– 0.7	
Seawater bony fish	0.3 – 1.3	0.1 – 2.7		24 – 66
Carp	56	6	0.2	6.5
Frog	11 – 21	68 – 78		
Alligator	67 – 81	0.17	up to 20	
Turtle	10 – 22	24 – 48	7 – 28	5.5
Chicken	15 – 20	9 – 11	59 – 70	7 – 9
Goat	3 – 19	29 – 76		4 – 23
Sheep	1.6– 18	51 – 83		3 – 22
Cattle	1.6– 20	27 – 60		6 – 13
Mouse	7.7	64 – 68	0.3	1.7
Primates	1.6– 1.8	54 – 84	0.1– 3.1	3.5– 6.4

Table 1.5.5: Urea in the Urine of Selected Mammals

For further figures on the composition of the urine, see Table 1.5.4.

From HALTENORTH 1977, modified

Species	Urea as percentage of urine
Blue whale	1.1
Cattle	1.1 –1.6
Common or Harbor seal	0.3 –3.0
Dog	0.7 –3.5
Elephant	0.18–0.45
Harbor porpoise	1.3 –2.4

Species	Urea as percentage of urine
Horse	0.8 –1.6
Human	1.8
Hyena	2.8 –3.4
Mouse	1.5 –3.2
Rabbit	0.55–0.9
Rat	1.8 –2.7
Sheep	0.6 –0.8
Tiger	2.9 –3.6

Table 1.5.6: Freezing Point Depression and Osmotic Pressure of Urine in Selected Animals

In some animals the value is dependent upon that of the external medium (seawater, freshwater); in these cases its freezing point depression is also provided. If this figure is not given, then the given values are independent of the external medium.

After HALTENORTH 1977, HEIDERMANNS 1957

Species	Freezing point depression Δ °C External medium	Urine	Osmotic pressure of urine (atm)	([hPa])
Cat		4.73	52.3	[52,993]
Cattle		1.34	16.2	[16,415]
Crayfish	0.02	0.09	1.1	[1,115]
Dog		3.29	39.8	[40,327]
Elephant		0.34	4.1	[4,154]
Freshwater bony fishes	0.02	0.10	1.2	[1,216]
Freshwater cartilaginous fishes	0.02	0.70	8.5	[8,613]
Frog	0.02	0.17	2.1	[2,128]
Goat		1.25	15.1	[15,300]
Guinea pig		0.91	11.0	[11,146]
Horse		0.32	3.9	[3,952]
Human		1.29	15.6	[15,807]
Lobster	1.85	1.89	22.9	[23,203]
Mitten crab	2.20	2.14	25.9	[26,243]
Mitten crab	0.02	1.23	14.9	[15,097]
Pig		1.70	20.6	[20,873]
Rabbit		0.92	11.1	[11,247]
Seawater cartilaginous fishes	1.85	1.92	23.2	[23,507]
Shore crab	1.00	1.50	18.2	[18,441]
Shore crab (*Carcinus*)	1.85	1.90	23.0	[23,305]

Table 1.5.7: Aquatic Salt Concentration Tolerance Limits of Selected Animals

Figures for salt given per mille (gram per liter).

Figures from ALTMAN and DITTMER 1966

Animal species	Salt ‰ (g/l)
Fish	
Carp	0– 10
Goldfish	0– 15
Plaice	7– 75
Rainbow trout	0– 35
Sticklebacks (Gasterosteidae)	0– 55
Invertebrates	
Brine shrimp	35–300
Bristleworm (*Nereis*)	0–139
Cockle	22– 60
Mitten crab	0– 47
Moon jelly	8– 35
Octopus (*Octopus*)	30– 35
Sand hopper (*Corophium*)	17– 72
Shore crab	27– 72
Shrimp (*Crangon*)	12– 35
Squid (*Loligo*)	31– 36

1.6 Heart, Circulation

Table 1.6.0: Numbers that Amaze

This table brings together some interesting figures from Sects. 1.6 Heart, Circulation, 1.7 Blood, and 1.8 Respiration. These numbers are cited again in the individual tables of these sections, where they are accompanied by references to the literature.

Weight of the heart			
	Blue whale	Absolute	598.4 kg
		In relation to BW	4.4 ‰
	Hummingbird	Absolute	0.09 g
		In relation to BW	23.9 ‰
Heart frequencies			
	Fruit fly (*Drosophila*)		235/min
	Shrew		1,320/min
	Elephant seal	(On the surface)	60/min
		While diving	4/min
Blood pressure (systolic/diastolic)			
	Giraffe		340/230 mmHg
	Canary		220/154 mmHg
	Mouse		147/106 mmHg
Number and size of red blood cells			
	Goat	18.5 million/mm³	3.5 μm × 3.5 μm
	Olm	36,000/mm³	58 μm × 34 μm
Coagulation times of blood			
	Lamprey		2–5 s
	Grass snake		27 min
Inner surface of the lung			
	Horse		500 m²
	Human		90 m²
	Fire salamander		5 cm²
Breaths per minute			
	Hummingbird		250
	Mouse		163
	Fin whale		0.5
Maximum diving times and depths of selected animals with lungs			
	Beaked whale	120 min	500 m
	Sperm whale	75 min	1,143 m
	Leatherback turtle		1,200 m

Table 1.6.1: Heart Weights of Selected Vertebrates (Absolute and Relative)

The relative weights are given as a per mille of body weight (BW); the absolute weights are in grams. The smaller relative heart weights of larger animals is related to their lower basal metabolic rate. See Table 1.4.1.

Data after BUDDENBROCK 1967, CZICSÀKY 1984, HALTENORTH 1977, DORST 1972, HESSE and DOFLEIN 1935, STURKIE 1976, ZISWILER 1976

Species	Relative heart weight (‰ of BW)	Absolute heart weight (g)
Fish		
Carp	1.5	1.62
Electric rays	0.6	
Pike	1.9	0.66
Salmon	2.0	10.72
Tiger shark	1.2	41.6
Trout	1.2	0.31
Tuna	3.1	616
Amphibians		
Axolotl	4.5	0.03
Bullfrog	3.2	1.65
Fire salamander	1.86	0.04
Green toad	8.05	
Marsh or Lake frog	1.8	0.14
Northern crested newt	1.96	
Tree frog	5.79	
Reptiles		
Alligator	2.6	137
Boa constrictor	3.1	5.64
Python	3.0	18.5
Western green lizard	3.8	0.91
Birds		
Blackbird	12.9	1.33
Budgerigar	13.2	
Carrion crow	10.7	
Chaffinch	13.2	0.3
Chick (just hatched)	9.0	
Chicken (leghorn)	6.4	
Common buzzard	7.5	4.81
Common kestrel	9.6	
Common raven	9.0	14.29

Species	Relative heart weight (‰ of BW)	Absolute heart weight (g)
Common swift	16.5	0.71
Domestic pigeon	14.0	
Duck	9.5	
Eagle owl	4.7	
Eurasian jay	9.9	1.52
European starling	16.2	0.93
Goose	8.0	
Great spotted woodpecker	13.4	1.04
Great tit	13.4	0.24
Herring gull	9.8	5.24
House sparrow	15.2	0.42
Hummingbird	23.9 (!)	0.09
Little owl	8.25	
Mallard	8.0	8.75
Northern lapwing	15.1	3.16
Ostrich	9.8	1,205
Pheasant	4.5	5.61
Stork	8.6	28.75
Swallow	14.1	0.32
Swan	10.3	67.6
Winter wren	16.2	0.14
Wood pigeon	10.63	5.32

Mammals

African elephant	3.9	19,500
Bison	6.6	6,600
Blue whale	4.4	598,400
Capuchin monkey	8.0	
Common house mouse	6.45	0.19
Dog	6.8	135
European rabbit	3.21	
Fin whale	5.0	380,000
Fox	9.0	63
Greyhound	12.9	
Guinea pig	4.0	4.8
Hare	9.42	
Harvest mouse	12.9	0.07
Hedgehog	4.0	4.8
Horse	9.0	4,000
Human	4.3	300
Jaguar	5.4	
Lion	3.0	750

(Cont.) →

Species	Relative heart weight (‰ of BW)	Absolute heart weight (g)
Mole	5.8	0.23
Myotis (bat)	10.45	0.22
Polar bear	5.8	2,900
Rat	3.62	1.45
Rhinoceros	6.3	
Roe deer	8.0	400
Shrew	14.0	0.11
Sperm whale	3.2	116,000
Squirrel	6.7	3.22
Weasel	16.7	0.75
Wood mouse	7.0	0.13

Table 1.6.2: Heart Rates of Selected Animals

Values given are only directly comparable with one another in birds and mammals. In poikilothermic animals they are strongly dependent on temperature. Heart rates in small homoiothermic animals reflect basal metabolic rates that are higher than in their larger counterparts. See also Table 1.4.1

After BUDDENBROCK 1967, HALTENORTH 1977, KOLB 1974, OPPENHEIMER and PINCUSSEN 1925, ZISWILER 1976

Species	Heartbeats per minute	Species	Heartbeats per minute
Invertebrates			
Amber snail	26	Pond mussel	4–6
Anopheles mosquito	110	Primitive insect (*Machilis*)	150–160
Aquatic sowbug	310	Privet hawk-moth (caterpillar)	39–82
Blue mussel	10–15	Privet hawk-moth (imago)	40–50
Cockroach	60–80	Shrimp	146
Cross spider (Araneus)	130–134	Small white (butterfly)	29
Dragonfly (*Anax*)	60	Stag beetle	16
Edible crab	90–120	Tunicate (*Ciona*)	17
Edible snail (*Helix pomatia*)	40–50	Woodlouse	180–200
Fruit fly	235		
Lobster	50		
Lugworm	13–22		
Migratory locust	80–130		
Octopus (*Octopus*)	33–40		

Species	Heartbeats per minute	Species	Heartbeats per minute
Vertebrates		Goat	70–80
Bat (*P. pipistrellus*)	660	Golden or Syrian hamster	350–500
Brown rat	260–450	Goldfish	36–40
Caiman	36–50	Goose	80
Capuchin monkey	160	Grass snake	23–40
Carp	40–80	Guinea pig	200–312
Cat	110–130	Hedgehog (hibernating)	18
Cat shark	65	Hedgehog (nonhibernating)	280–320
Cattle	45–50	Horse	32–44
Chicken	330–375	Human	60–90
Common buzzard	206–351	Lion	40
Common frog	35–40	Mouse	450–550
Common pipistrelle	972	Ostrich	60–70
Common swift	700	Panther	60
Common toad	40–50	Pig	60–80
Crossed viper	40	Pike	30–42
Crow	380	Rabbit	150–280
Crow	300–352	Rattlesnake	44–47
Dog	60–180	Rock dove	150–250
Dolphin	84–140	Sand lizard	60–70
Duck	229–420	Sheep	60–80
Eel	10–15	Shrew	500–1,320 (!)
Electric rays	16–50	Slowworm	64
Elephant	22–28	Sparrow	745–850
Elephant seal		Stork	270
on the water surface	60	Turkey	93
while diving	4	Western green lizard	60–66
Fox	100	Whale	15–16
Giant tortoise	30–40		
Giraffe	66		

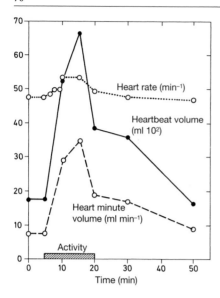

Fig. 1.6.2: Dependency of cardiac activity (heart minute volume, heartbeat volume, and heart rate) on temperature and activity in rainbow trout. See also Tables 1.6.2 and 1.6.3. From PENZLIN 1980, p. 251

Table 1.6.3: Heart Minute Volumes of Selected Animals

The heart minute volume (HMV) given refers to the amount of blood in cubic centimeters moved by the heart of the given animal in 1 min. Strong physical exertion can raise this figure considerably, in humans up to 30 l. It is calculated by multiplying the heart rate (Table 1.6.2) by the heartbeat volume.

Figures from BUDDENBROCK 1967

Species	HMV (cm³/min)		
Cattle	45,800		
Chicken	400		
Dog	2,000–	3,000	
Frog	20–	30	
Horse	18,000–24,000		
Human	5,500–	6,000	
Migratory locust			0.015
Octopus (*Octopus*)	57–	81	
Pig	4,500		
Turtle		148	

Table 1.6.4: Blood Pressure of Selected Animals

Figures represent systolic and diastolic blood pressure, respectively, separated by a slash.
Single numbers represent mean values.

After ALTMAN and DITTMER 1974, BUDDENBROCK 1967, HALTENORTH 1977, KOLB 1962

Species	Blood pressure (mmHg)	(in hPa)
Canary	220/154	(293/205)
Cat	155/100	(207/133)
Cattle	134/88	(179/117)
Chicken	150/120	(200/160)
Chimpanzee	136/80 (!)	(181/107)
Common or Harbor seal	150/105	(200/140)
Crocodile	50	(67)
Cuttlefish	44/22	(59/29)
Dog	156/100	(208/133)
Duck	180/134	(240/179)
Eel	35– 40	(47–53)
Electric rays	16/14	(21/19)
European starling	180/130	(240/173)
Frog	32/19	(43/25)
Giraffe	340/230	(453/307)
Goat	120/84	(160/112)
Golden or Syrian hamster	170/120	(227/160)
Grass snake	89	(119)
Guinea pig	77/47	(103/63)
Horse	190/120	(253/160)
Human	120/80	(160/107)
Kangaroo	122/79	(163/105)
Lobster	8	(11)
Mouse	147/106	(196/141)
Pig	169/108	(225/144)
Pike	35.5	(47)
Rabbit	110/80	(147/107)
Rat	145/105	(193/140)
Rock dove	187/138	(249/184)
Salamander	22/12	(29/16)
Salmon	75	(100)
Shark	32/16	(43/21)
Sheep	103/85	(137/113)
Slider turtle	30/26	(40/35)
Sparrow	180/140	(240/187)
Trout	39/34	(52/45)

1.7 Blood

Table 1.7.1: Blood Volumes of Vertebrates

Data given as percentage of body weight (BW). Considerable variations are possible in some cases.

Data after BERTELSMANN 1979, OPPENHEIMER and PINCUSSEN 1925, PROSSER 1973

Species	Blood volume as percentage of BW	
	Mean value	Limits
Bat (*Myotis*)	13.0	
Blue whale	6.5	
Cat	5.6	
Cattle		5.2– 5.7
Chicken	7.0	
Common frog	5.6	4.7– 6.3
Crocodile	15.4	
Dog	8.6	2.3– 8.7
Duck	10.2	
Eel	2.9	
Fat dormouse	5.8	5.3– 6.0
Goat	7.3	
Grass snake	7.7	
Guinea pig	7.5	
Hare	7.5	6.4– 8.1
Hedgehog	8.0	
Horse	7.6	
Human ♂		6.5– 7.1
Human ♀		7.1– 7.8
Lizard	5.8	4.7– 7.0
Mouse	5.8	4.6– 7.0
Mudpuppy	2.9	
Pig	4.6	2.3– 8.7
Rat	7.5	
Rock dove	7.8	
Rooster	9.0	
Salamander	6.1	5.5– 6.8
Salmon	2.8	
Sea lamprey	4.9	
Sharks	7.0	
Sheep	8.1	6.6–10.4
Slowworm	5.2	3.8– 7.6
Turtle	9.1	

Table 1.7.2: Number and Size of Red Blood Cells in Selected Vertebrates

The number is in millions per cubic millimeter; size is given in micrometers.

After BUDDENBROCK 1967, HALTENORTH 1977, HEIDERMANNS 1957, HESSE and DOFLEIN 1935, KOLB 1974, *Kosmos-Lexikon* 1953, PORTMANN 1959, STURKIE 1976

Species	Size (μm)	Number (1,000,000/mm³)
Carp	11.0 × 8.2	1.40
Carrion crow	12.0 × 7.0	2.49
Cat	4.5 × 4.5	8.22
Chaffinch	12.4 × 7.5	3.66
Chimpanzee	7.4 × 7.4	5.10
Common frog	23.0 × 16.0	0.49
Common toad	20.5 × 13.3	0.38
Dog	7.0 × 7.0	6.65
Edible frog	23.0 × 16.0	0.60
Eel	15.0 × 12.0	1.1
Electric rays	27.0 × 20.0	0.14
Elephant	9.0 × 9.0	2.02
Fire salamander	43.0 × 25.0	0.09
Goat	3.5 × 3.5	18.50 (!)
Golden or Syrian hamster	5.6 × 5.6	6.90
Grass snake	18.0 × 11.0	0.97
Guinea pig	7.4 × 7.4	5.60
Horse ("coldblood")		6.90
Horse (thoroughbred)		8.90
Human ♀	7.5 × 7.5	4.50
Human ♂	7.5 × 7.5	5.00
Lamprey	15.0 × 15.0	0.13
Llama	7.6 × 4.4	13.19
Mouse	6.0 × 6.0	9.50
Mudpuppy	58.0 × 34.0 (!)	0.036
Musk deer	2.5 × 2.5 (!)	
Natterjack toad	20.7 × 14.4	0.65
Northern crested newt	31.0 × 22.0	0.164
Ostrich	14.0 × 9.0	1.62
Pig	6.0 × 6.0	6.96
Rat	7.5 × 7.5	9.00
Rock dove	14.0 × 7.0	3.00
Sand lizard	16.0 × 10.0	1.42
Slowworm	18.0 × 9.0	1.52
Sole	12.0 × 9.0	2.00
Walrus	9.2 × 9.2	

Table 1.7.3: Number of Platelets of Selected Animals

Figures given in thousands per cubic centimeter. For the number of red blood cells, see Table 1.7.2; for the lifespan of platelets, see Table 1.7.5.

After HALTENORTH 1977, STURKIE 1976

Species	Number of platelets (1,000s per mm³)
Cat	164– 500
Cattle	542– 975
Chicken	30.4
Dog	188– 960
Golden or Syrian hamster	160– 516
Guinea pig	525– 900
Horse	249– 461
Human	200– 500
Mouse	246– 339
Ostrich	10.5
Pig	296– 616
Rabbit	170–1.120
Rat	702– 796
Rhesus monkey	155– 424
Rock dove	13.0
Sheep	284– 659

Table 1.7.4: Coagulation Times of Blood in Vertebrates

Figures from BERTELSMANN 1979. HEIDERMANNS 1957. KOLB 1974. MEYER 1964

Species	Coagulation time	
Cat	1.0– 3.0	min
Cattle	6.5	min
Chicken	4.5	min
Cod (Baltic Sea)	3.0	min
Dog	4.5	min
Edible frog	5.0	min
Goat	2.5	min
Grass snake	27.0	min
Horse	11.5	min
Human	5.0– 6.0	min
Lamprey	2.0– 5.0	s
Perch	20.0	s
Pig	3.5	min
Rabbit	4.0	min
Rock dove	1.5	min
Sheep	2.5	min

Table 1.7.5: Lifespan of Blood Cells

In white blood cells, both granulocytes and lymphocytes, there is great variation among the individual forms

Figures from ALTMANN and DITTMER 1972, BUDDENBROCK 1967, PLATT 1976

Species	Lifespan	
Erythrocytes		
Cat	68– 77	days
Chicken	35	days
Dog	90–135	days
Duck	42	days
Goat	106–125	days
Guinea pig	80– 90	days
Horse	140–150	days
Human	108–150	days
Mouse	20– 45	days
Pig	62– 71	days
Rabbit	45– 68	days
Rat	45– 68	days
Rock dove	105	days
Sheep	70–153	days
Turtle	500	days
Granulocytes		
Human	30– 40	days
Lymphocytes		
Human		
Effector cells	10	days
Memory cells	>10	years
Rat	14 days – 9	months
Platelets		
Cat	2– 4	days
Cattle	10	days
Dog	7– 9	days
Guinea pig	5	days
Human	8– 14	days
Mouse	4	days
Rabbit	6	days
Rat	4– 5	days

Table 1.7.6: Hematocrit Levels of Selected Vertebrates

The hematocrit is the percentage of total blood volume occupied by red blood cells.

After BUDDENBROCK 1967, HALTENORTH 1977, KOLB 1974, STURKIE 1970

Alligator	30.0–31.0	Goose	44.0
Bonito (fish)	37.1	Guinea pig	42.0
Carrion crow	53.8	Horse ("coldblood")	35.0
Cat	40.0	Horse (thoroughbred)	42.0
Cattle	40.0	Human	45.0
Chicken	34.5–37.5	Mouse	41.5
Chimpanzee	41.5	Opossum	35.3
Colubrid [snake] (*Natrix*)	21.9	Pig	41.5
Common frog	41.8	Rabbit	41.5
Common or Harbor seal	48.0	Rat	46.0
Dog	45.5	Rock dove	58.5
Dolphin	51.7	Sheep	32.0
Duck	39.5	Slider turtle	31.5
Goat	34.0	Sparrow	37.0–52.0
Golden or Syrian hamster	46.0		

Table 1.7.7: Hemoglobin Levels in the Blood of Selected Animals

Data in grams hemoglobin (Hb) per 100 cm^3 blood. For more detailed figures on humans, see Table 4.3.11.

Data after BUDDENBROCK 1967, HALTENORTH 1977, KOLB 1974, MEYER 1964, STURKIE 1976, WITTKE 1972

Bat	15.0	Great ramshorn snail	1.5
Carp	12.2	Guinea pig	15.9
Carrion crow	16.4	Horse	12.6
Cat	10.2	Human ♂	12.0–16.0
Cat shark	4.8	Human ♀	14.0-18.0
Cattle	16.0	Hummingbird	18.0
Chaffinch	13.9	Long-beaked echidna	16.7
Chicken	11.2	Mouse	12.5
Common frog	13.0	Pheasant	18.9
Dog	16.8	Pig	13.3
Earthworm	3.8	Rabbit	11.8
Eel	10.9	Reptiles	6.9
Fox	11.0	Rock dove	13.7–16.0
Goat	10.6	Sheep	12.5
Golden or Syrian hamster	15.0	Sparrow	14.0
Goose	12.7	Thornback ray	2.3

Table 1.7.8: Molecular Weight of the Hemoglobin of Selected Animals

In the case of the hemoglobin of invertebrates, which is not stored in the blood corpusles, just a relatively small number of larger molecules allow for a high level of oxygen capacity (see also Table 1.7.9) without a marked increase in the osmotic pressure of the blood or hemolymph. On this, cf. also Table 5.3.

From HEIDERMANNS 1957

Species	Molecular weight
Earthworm	3,000,000
Great ramshorn snail (*Planorbis*)	1,540,000
Lamprey	19,100
Lugworm (*Arenicola*)	3,000,000
Midge larvae (*Chironomus*)	31,500
Mussel (*Arca*)	33,000
Vertebrates	68,000
Water flea	400,000

Table 1.7.9: Oxygen Capacity of the Blood (or Hemolymph) of Selected Animals

All figures are given in cubic centimeters oxygen per hundred cubic centimeters blood.

Figures from BUDDENBROCK 1967, FLOREY 1970, HALTENORTH 1977, KOLB 1974

Species	Oxygen capacity (cm³ O₂/100 cm³ blood)
Invertebrates	
Blue mussel	0.32
Earthworm	5.10– 5.40
Edible crab	3.44
Edible snail (*Helix pomatia*)	1.10– 2.20
Great ramshorn snail	1.20– 2.80
Lobster	1.30
Lugworm	5.00
Midge larva	5.40–11.60
Octopus (*Octopus*)	3.10
Roundworm	3.00
Scorpion	1.80
Shore crab	1.14– 1.16
Starfish	0.46

(Cont.) →

Species	Oxygen capacity (cm³ O₂/100 cm³ blood)

Vertebrates

Adelie penguin	22.40
Carp	12.50
Cat	13.10–14.30
Cattle	16.80
Chicken	10.00
Chimpanzee	17.40
Crocodile	8.00–10.00
Dog	19.80
Edible frog	9.80–13.30
Electric eel	19.70
Goat	14.20
Golden or Syrian hamster	20.80
Goldfish	10.70
Guinea pig	21.30
Harbor porpoise	18.20
Hedgehog	15.30
Horse	16.70
Human	21.30
Lamprey	1.20
Lungfish	6.80
Mackerel	15.70
Mouse	17.30
Orca	21.50
Pig	17.80
Rabbit	16.00
Rock dove	20.00
Shark	4.50
Sperm whale	29.90
Turtle	6.60–10.80

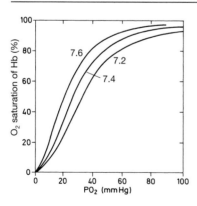

Fig. 1.7.9: Influence of the pH value of blood on oxygen saturation of hemoglobin. See also Table 1.7.9 From Remane, Storch and Welsch 1985

Table 1.7.10: Blood Sugar Levels in Selected Vertebrates

Figures in milligrams sugar per 100 cm³ blood.

Figures from BUDDENBROCK 1956, KOLB 1974

Species	Blood sugar level (mg/100 cm³)
Alligator	59– 97
Black redstart	347
Canary	236
Cat shark	27– 39
Cattle	40– 70
Chicken	130–260
Common swift	305
Devil anglerfish (bony fish)	5– 25
Dog	60– 90
Duck	148
Electric rays	13– 77
Frog (*Rana*)	30– 75
Goat	45– 60
Horse	55– 95
Houndshark	65–137
Human	60–100
Ostrich	164
Pig	45– 75
Rabbit	97–109
Rat	92–106
Rock dove	190
Sheep	30– 50
Sparrow	288
Turtle (*Emys*)	58– 99

Table 1.7.11: Protein Levels of Bodily Fluids
(Blood, Hemolymph, Coelom Fluid) of Selected Animals

Figures in milligrams protein per cubic centimeter. 10 mg/cm^3 = 1 %. The figures represent mean values.

Figures from BUDDENBROCK 1967, PROSSER 1973

Species	Protein level (mg/cm^3)
Alligator	51.7
Blue mussel	1.5
Bony fish (common whitefish)	41.9
Bumblebee	50.0
Cattle	50.2
Chicken	31.8
Cuttlefish (*Sepia*)	49.8
Dog	58.0–70.0
Dolphin	75.0
Edible snail (*Helix pomatia*)	24.0–33.0
Frog (*Rana*)	24.0
Goose	30.0
Horse	75.0
Human	75.0
Jellyfish	0.4
Lamprey	36.0
Lizard	45.7
Lobster	40.0
Lugworm (*Arenicola*)	0.2
Pond mussel	3.0
Rabbit	54.4
Rat	59.0
Ray	41.2
Salamander	21.6
Sea cucumber	0.7
Sea mouse (bristleworm)	0.9
Shark	30.0
Shore crab	40.0–60.0
Silkworm, caterpillar	10.0
Silkworm, imago	60.0
Slowworm	40.3
Starfish	0.6
Walking stick	10.0

Table 1.7.12: Freezing Point Depression and Osmolarity of the Blood of Selected Animals

Osmolarity can be calculated from the easily measurable extent of freezing point depression of a liquid, since a molar solution exhibits a freezing point depression of 1.85 °C and an osmotic pressure of 22.4 atm (22,697 hPa). See also Table 1.7.13.

Data after HALTENORTH 1977, HANKE et al. 1977, HEIDERMANNS 1957

Species	Freezing point depression (Δ °C)	Osmolarity (mOsm)	Osmotic pressure (Atm)	([hPa])
Blue mussel	2.26	1,216	27.3	[27,662]
Carp	0.50	270	6.0	[6,080]
Cat	0.63	339	7.6	[7,701]
Cat shark	2.22	1,195	26.8	[27,155]
Cattle	0.59	318	7.2	[7,295]
Cod (Baltic Sea)	0.76	409	9.2	[9,222]
Crayfish	0.80	431	9.7	[9,829]
Dog	0.60	323	7.3	[7,397]
Earthworm	0.45	242	5.4	[5,472]
Edible frog	0.40	215	4.8	[4,864]
Eel	0.62	334	7.5	[7,599]
Horse	0.56	302	6.8	[6,890]
Human	0.56	301	6.8	[6,890]
Lobster	1.89	1,017	22.8	[23,102]
Pig	0.61	329	7.4	[7,498]
Pinniped	0.70	377	8.4	[8,511]
Plaice	0.79	425	9.5	[9,626]
Pond mussel	0.09	48	1.1	[1,115]
Praying mantis	0.89	479	10.7	[10,842]
Rabbit	0.64	345	7.8	[7,903]
Sheep	0.62	334	7.5	[7,599]
Shore crab	1.97	1,060	23.8	[24,115]

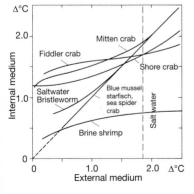

Fig. 1.7.12: Dependency of freezing point depression in blood upon that in external medium. See also Table 1.7.12.
From PENZLIN 1980, p. 311

Table 1.7.13: Freezing Point Depression und Osmotic Pressure of Hemolymphs of Arthropods

A molar solution has a freezing point depression of 1.85 °C and an osmotic pressure of 22.4 atm (22,697 hPa). See also Table 1.7.12.

After BUDDENBROCK 1967

Species	Freezing point depression (Δ °C)	Osmotic pressure (atm)	([hPa])
Drosophila larva	0.700	8.5	[8,613]
Ground beetle (*Carabus*)	0.940	11.4	[11,551]
House cricket	0.720	8.7	[8,815]
Meal moth larva	1.120	13.6	[13,780]
Mosquito larva	0.650	7.9	[8,005]
Praying mantis	0.885	10.7	[10,842]
Scorpion	1.125	13.6	[13,780]
Spider	0.894	10.8	[10,943]

Table 1.7.14: Physiological Salt Solutions (Ringer's, Tyrode's, Locke's Solutions) for Selected Animals

Figures in grams per liter distilled water. Possible shares of water of crystallization of the substances was not taken into account.

After BAUER et al. 1974, HEIDERMANNS 1957

Solution	NaCl	KCl	$CaCl_2$	$MgCl_2$	NaH_2PO_4	$NaHCO_3$	Glucose
Ringer's earthworm	6.00	0.12	0.20	–	0.10	–	–
Ringer's insect	10.93	1.57	0.83	0.17	–	–	
Ringer's insect	7.60	0.75	0.22	0.19	0.37	0.48	–
Ringer's isopod	14.70	0.70	1.67	0.95	–	–	–
Ringer's frog (summer)	6.00	0.05	0.29	–	–	0.30	–
Ringer's frog (winter)	6.00	0.10	0.15	–	–	0.30	–
Locke's frog	6.00	0.30	0.20	–	–	0.20	–
Ringer's mammal	8.70	0.20	0.20	–	–	0.1–1.0	–
Tyrode's mammal	8.00	0.20	0.20	0.10	0.05	1.0	1.0
Locke's mammal	9.20	0.42	0.24	–	–	0.15	–

1.8 Respiration

Table 1.8.0: Numbers that Amaze

This table brings together some interesting numbers on respiration in animals, which are cited again in the individual tables below.

Interior lung surface		
	in humans	90 m^2
	in horses	500 m^2
Breaths per minute		
	in hummingbirds	250
	in mice	163
	in Eurasian siskins	114
Air volume inhaled and exhaled in one breath		
	in orcas	46,200 cm^3
	in horses	7,500 cm^3
	by comparison: in humans	500 cm^3
Maximum diving time		
	in beaked whales	120 min
	in leatherback turtles	120 min
	in bowhead whales	80 min
Maximum diving depth		
	in sperm whales	1,143 m
	in leatherback turtles	1,200 m
Oxygen consumption during exertion per kilogram body weight		
	in mice	28 g
	in humans	5.6 g

Table 1.8.1: Relative Lung Weight in Selected Vertebrates

Values vary greatly. For this reason, a range is given for humans. The figures given for animals are mean values. Figures represent percentage of body weight.

Data after CsicsÀky 1984, Haltenorth 1977

Species	%
Mammals	
African elephant	2.08
Beluga whale	2.70
Cat	1.04
Cattle	0.72
Chimpanzee	1.36
Common vole	1.70
Dog	0.94
Golden or Syrian hamster	0.46
Guinea pig	1.18
Hippopotamus	0.84
Horse	0.70
Human	0.73–3.13
Lion	2.12
Rat	0.79
Rhesus monkey	1.89
Walrus	1.36
Wolf	3.56
Birds	
Budgerigar	0.73
Carrion crow	1.19
Common buzzard	0.64
House sparrow	1.35
Hummingbird	1.37
Mallard	0.83
Ostrich	2.36
Pheasant	0.34
Rock dove	1.06
Stork	0.81
Swan	0.65

Table 1.8.2: Interior Lung Surface of Selected Animals

Figures represent mean values. Within a given class of vertebrates and among animals of the same size, those animals that move quickly are equipped with a larger lung surface.

Values after ALTMAN and DITTMER 1974, HALTENORTH 1977, KRUMBIEGEL 1953, ZISWILER 1976

Species	Interior lung surface (m²)
Alligator	2.0
Bat (myotis)	0.5
Cat	20.0
Common or Harbor seal	65.0
Common toad	0.008
Dog	90.0
Dolphin	43.0
Fire salamander	0.0005
Horse	500.0
Human	90.0
Lungfish	0.016
Mouse	0.12
Rabbit	5.9
Rat	0.56
Sloth	5.0

Table 1.8.3: Breathing Rate and Single-Breath Air Volume of Selected Vertebrates

Data after BERTELSMANN 1979, Enzyklopädie 1979, KRUMBIEGEL 1953, MEYER 1964, OPPENHEIMER and PINCUSSEN 1925, PENZLIN 1977, ZISWILER 1976

Species	Breaths/min	Volume (cm³)
Bat	50	
Buzzard/Hawk	20	
Capuchin monkey	46	
Carrion crow	25	
Cat	30	34
Cat shark	52	
Cattle	30	3,400
Chicken	27	31
Chimpanzee	19	
Common swift	90	
Dog	18	320
Duck	42	30
Electric rays	60	
Elephant	6	

(Cont.) →

Species	Breaths/min	Volume (cm³)
Eurasian siskin	114	
European conger	48	
Fin whale	0.5	
Fox	24	
Giraffe	9	3,400
Golden or Syrian hamster	74	0.83
Gorilla	19	
Greater weever fish	26	
Guinea pig	90	1.75
Harbor porpoise	1.1	9,000
Hedgehog. awake	20	
Hedgehog. hibernating	5	
Horse	10	7,500
Human	11	500
Hummingbird	250 (!)	
Lion	10	
Lungfish (*Neoceratodus*)	28	15
Moray eel	26	
Mouse	163	0.15
Orca	1.1	46,200
Ostrich	3	
Painted comber (fish)	56	
Rabbit	37	21
Rat	85	0.86
Rock dove	30	4.7
Sheep	20	362
Shrew	120	
Sparrow	90	
Stork	8	
Wolf	14	

Table 1.8.4: Oxygen Consumption of Selected Animals

Figures in grams oxygen per kilogram body weight and hour. One gram of oxygen (O_2) corresponds to approx. 700 cm³. See also Table 1.4 and the comments found there on the relationship between oxygen consumption and body size.
Listed in systematic order.

Figures from SCHLIEPER 1952, VANGEROW 1975

Taxon, species	Body weight		Oxygen consumption (g O_2 per kg and h)
Paramecium	0.001	mg	0.7
Radiolaria (Protozoa)	0.1	mg	0.1
Cnidaria	10–100	g	0.004–0.02
Mussels	25–250	g	0.002–0.03
Snails	20–50	g	0.01 –0.1
Cephalopods	10	g	0.6 –0.32
Segmented worms (annelids)	1–20	g	0.03 –0.7
Crustaceans	5–500	g	0.04 –0.28
Echinoderms	10–600	g	0.018–0.04
Carp	200	g	0.08 –0.3
Pike	200	g	0.49
Frog	30	g	0.03 –0.09
			max. 0.63
Reptiles	30	g	0.02 –0.03
Chicken	2	kg	0.79 –1.1
Mouse (resting)	20	g	3.5
(running)			28
Rat (resting)	110	g	2.4
Dog (resting)	9	kg	0.72
Horse	400	kg	0.3
Human (resting)	70	kg	0.28
(working)			5.6

Table 1.8.5: Oxygen Consumption During Hibernation of Selected European Mammals

Figures represent oxygen consumption in cubic centimeters oxygen (O_2) per kilogram body weight (BW) and hour. For figures for nonhibernating animals, see Table 1.8.4. On hibernation, see also Table 1.4.5.

Data after ALTMAN and DITTMER 1972

Species	Oxygen consumption (cm³ O_2 per kg BW and per hour)
Common pipistrelle	53
European ground squirrel (suslik)	15

(Cont.) →

Species	Oxygen consumption (cm³ O₂ per kg BW and per hour)
Fat dormouse	17–29
Giant noctule	30
Hamster	32
Hazel dormouse	40
Hedgehog	28
Long-eared bat	69
Marmot	18
Myotis (bat)	20

Table 1.8.6: Maximum Diving Depth and Diving Time of Air-Breathing Vertebrates

Figures from BRINK 1975, FRÄDRICH and FRÄDRICH 1973, GRZIMEK 1970, KRUMBIEGEL 1953, *Nat. Rdsch.*, NIETHAMMER 1979, PENZLIN 1970, SCHÄFER 1968, WOOD 1982, ZISWILER 1976

Species	Maximum diving time	Maximum depth
Alligator	120 min	
Beaked whale	120 min (!)	500 m
Beaver	15 min	
Blue whale	30 min	200 m
Bowhead whale	80 min	1,000 m
Common or Harbor seal	15 min	
Duck	15 min	
Elephant seal	30 min	100 m
Gray seal	18 min	146 m
Great cormorant	5 min	40 m
Great crested grebe	56 s	40 m
Guillemot	12 min	
Harbor porpoise		170 m
Hippopotamus	19 min	
Human	2 min	72 m
Leatherback turtle		1,200 m (!)
Loon	15 min	75 m
Muskrat	12 min	
Northern gannet		10 m
Penguins		60 m
Platypus	12 min	
Polar bear	2 min	
Sea otter		50 m
Sirenia ("sea cows")	10 min	
Sperm whale	75 min	1,143 m
Walrus	10 min	30 m
Weddell seal	43 min	600 m

1.9 Reproduction, Development

Table 1.9.0: Numbers that Amaze

This table brings together some interesting numbers from this section, which are also cited in individual tables, where they are accompanied by references to the literature.

Egg size and weight		
	Hummingbird	13 × 8 mm/0.2 g
	Ostrich	150 × 130 mm/1,500 g
	Foss. Madagascar's elephant bird	340 ×250 mm/10,000 g
	Python	120 × 60 mm
	Sea or shark catfish	22 × 22 mm
Sexual maturity begins		
	in Norway lemming ♀	at the age of 19 days
	in tuatara	at the age of 20 years
	in rhinoceros	at the age of 20 years
Incubation and nesting time lasts, respectively,		
	in royal albatross	80 days, 243 days
	in nightingale	13 days, 12 days
Pregnancy lasts		
	in opossum	13 days
	in elephant	up to 660 days
Relative birth weight (as a proportion of that of the mother)		
	in kangaroo	0.0003 %
	in bats	up to 40 %
Maximum number of eggs per clutch		
	in turbot	9,000,000
	in green toad	12,000
	in leatherback turtle	100
	in gray partridge	18
Diploid chromosome number		
	in lamprey	174
	in roundworm	2

Table 1.9.1: Size of Egg Cells in Selected Animals (Excluding Birds)

No figures for birds; for these, see Table 1.9.13. For fish, see Table 1.9.9. For insect eggs, the length is given

Data after *Lexikon der Biologie*, OPPENHEIMER and PINCUSSEN 1927, SLIJPER 1967, STARCK 1975, SPECTOR 1956

Species	Maximum egg diameter	
African clawed frog	1.5	mm
Alligator	50	mm
Bee	1.75	mm
Cat	130	μm
Cattle	150	μm
Coelacanth	100	mm
Colorado potato beetle	1.75	mm
Common vole	60	μm
Dog	140	μm
Frog	500	μm
Fruit fly	500	μm
Guinea pig	100	μm
Horse	140	μm
House cricket	2.5	mm
Human	100	μm
Lamprey	2	mm
Lancelet	110	μm
Lungfish	3.75	mm
Moon jelly	190	μm
Pig	140	μm
Pike	1	mm
Platypus	4	mm
Python	60 × 120	mm
Quoll	240	μm
Rabbit	130	μm
Rat	70	μm
Rhesus monkey	120	μm
Roundworm	45	μm
Sea urchin	70	μm
Sheep	180	μm
Whale	130	μm

Table 1.9.2: Dimensions of Sperm of Selected Animals

Figures from DAVID 1977

Species	Overall length (µm)	Length of head (µm)
Cat	54.0	4.0
Cattle	73.2	8.8
Guinea pig	93.0	13.0
Hedgehog	85.0	5.0
Human	58.0– 67.0	4.5– 5.5
Mouse	100.0	7.0
Rat	170.0–200.0	10.0

Table 1.9.3: Chromosome Sets of Selected Animals

Diploid values, numbers of chromosomes (*n*). For plants, see Table 2.2.9.

After ALTMANN and DITTMER 1972, BRESCH and HAUSMANN 1972, DAUMER 1977, Kosmos 1953, *Lexikon der Biologie*, NAGL 1976, NIGON and LUEKEN 1976, SPECTOR 1956, STARCK 1974

Species	Chromosomes (*n*)	Species	Chromosomes (*n*)
Invertebrates		Liver fluke	12
Bee	32	Migratory locust ♂	23
Cockroach ♂	47	Moon jelly	20
Cuttlefish (*Sepia*)	12	Mosquito (*Culex*)	6
Cyclops (water flea)	14	Praying mantis ♂	27
Dragonfly (*Aeschna*)	26	Roundworm	2
Earthworm	32	Scorpion (*Buthus*)	14
Edible snail (*Helix pomatia*)	54	Sea urchin (*Paracentrotus*)	36
Firebug	24	Silkworm	20
Fruit fly (*Drosophila*)	8	Silverfish	34
Great diving beetle	38	Small white (butterfly)	30
Head louse	12	Spider (*Aranea*)	14
Heliozoa	44	Sponge (*Sycon*)	26
House spider ♂	43	Starfish	36
Housefly	12	Stentor protozoan	28
Hydra	32	Water flea	20
Ixodes ricinus tick	28	White-lipped banded snail	48
Lancelet	24	Woodlouse	56
Leech (*Glossosiphonia*)	26	Yellow mealworm beetle	20

(Cont.) →

Species	Chromosomes (n)	Species	Chromosomes (n)
Vertebrates		Chicken	78
Fish		Duck	80
		Egret	68
Carp	104	Greylag goose	80
Cat shark	24	House sparrow	76
Goldfish	94	Rock dove	80
Guppy	48		
Lamprey	174		
Pike	18		
Swordtail	48	**Mammals**	
		Baboon	42
Amphibians		Bat (*Myotis*)	44
Axolotl	28	Capuchin monkey	54
Common Eurasian spadefoot toad	26	Cattle	60
Common frog	26	Chimpanzee	48
Common toad	22	Common house mouse	40
Edible frog	26	Dog	78
Fire salamander	24	Gibbon	52
Green toad	22	Golden or Syrian hamster	44
Midwife toad	36	Gorilla	48
Natterjack toad	22	Guinea pig	64
Northern crested newt	24	Hare	48
Tree frog	24	Hedgehog	48
		Horse	64
Reptiles		House cat	38
Alligator	32	Human	46
Asp viper	42	Kangaroo	22
Crossed viper	36	Macaque	44
European pond turtle	50	Mole	34
Hawksbill turtle	58	Orangutan	44
Sand lizard	38	Pig	38
Slowworm	44	Rabbit	44
		Rat	42
Birds		Rhesus monkey	42
Blackbird	80	Sheep	54
Canary	80	Tarsier	80

Table 1.9.4: Age of Sexual Maturity in Selected Animals

For a comparison of time of sexual maturity with lifespan in mammals, see Table 1.9.19.

Data after ALTMAN and DITTMER 1972, BOGNER and RITTER 1976, FRÄDRICH and FRÄDRICH 1973, GRZIMEK 1970, KRUMBIEGEL 1953, MEYER 1964, MICHEL 1986, SPECTOR 1956, ZISWILER 1976

Species	Age of sexual maturity	Species	Age of sexual maturity
Anura	3–4 years	Guinea pig	55–70 days
Baboon	4– 8 years	Hare	6 months
Bat	1 year	Hedgehog	2 years
Beaver	2 years	Herring	3– 7 months
Beech marten	2 years	Hippopotamus	2.5 years
Birds of prey	3– 6 years	Horse	3– 4 years
Bison	3 years	Human	12–18 years
Blue whale	4 years	Lion	3–5 years
Brown bear	5 years	Lizards	2– 3 years
Camel	4 years	Lobster	5 years
Carp	4– 5 years	Macaque	2 years
Cat	1 year	Mole	2 years
Cattle	1.5– 2 years	Mouse	2– 6 months
Chickens	1– 2 years	Norway lemming ♀	19 days (!)
Chimpanzee	8–10 years	Opossum	8 months
Common frog	3 years	Orangutan	10 years
Common or Harbor seal	4 years	Ostrich	3– 4 years
Crocodile	3–6 years	Platypus	2 years
Deer	1.5 years	Polar bear	2– 3 years
Dog	1 year	Rabbit	5– 9 months
Domestic pig	7 months	Rat	2– 3 months
Ducks	1– 2 years	Rhinoceros	20 years (!)
Edible snail (*Helix pomatia*)	2– 3 years	Roe deer	14 months
Elephants	12–15 years	Salamander	3– 4 years
European pond turtle	12 years	Sheep	1– 2 years
Fox	10 months	Shrew (*Sorex*)	1– 2 years
Gibbon	8–10 years	Songbirds	1 year
Giraffe	6– 7 years	Tiger	4– 5 years
Goat	1– 2 years	Trout	2– 4 years
Golden or Syrian hamster	2 months	Tuatara	20 years (!)
Gorilla ♂	6– 7 years	Wild boar	1.5 years
Gorilla ♀	9–10 years	Wolf	3 years

Table 1.9.5: Cell Division Time in Protozoa

For figures on cell division in higher organisms, see Table 5.7.

Data after OPPENHEIMER and PINCUSSEN 1930

Species	Cell division time
Amoebacristalligera	3 min
Amoeba vespertilio	1 h
Opalina ranarum	40–50 min
Paramecium aurelia	ca. 2 h
Paramecium caudatum	ca. 1 h
Pelomyxia binucleata	7.5 h
Stentor polymorphus	ca. 2 h
Stylonychia mytilus	1–2 h
Vorticella	30 min

Table 1.9.6: Reproduction and Development of Selected Insects

Figures from ALTMAN and DITTMER 1972, SPECTOR 1956

Species	Eggs per female	Duration of developmental stage (days)				Genera-tions
		Egg	Larva	Pupa	Imago	
Black bean aphid		90–120	10	–	7– 20	
Blowfly	100– 500	1– 2	4– 15	3–13	10– 40	2– 4
Cockroach (*Blatella germanica*)	218– 267	17	40– 42	–	80–232	1– 4
Cockroach (*Periplaneta americana*)	200–1,000	32– 58	285–616	–	102–558	£ 1
Fruit fly (*Drosophila*)	100	<1	3– 11	2– 8	14	5– 6
Grasshopper	300– 400	90–120	6– 40	–	> 30	1
Green-veined white	200– 500	7	14	7–14		3– 6
Honeybee		3	8	9	35– 40	
House cricket	728	56– 84	210–231	–	50– 90	1
Housefly	75– 720	1– 3	4– 10	4–18	10– 50	4–18
Louse	50– 300	5– 21	7– 10	–	10– 30	10–12
Meal moth	116– 700	>3	40	5– 7	3– 4	> 6
Mosquito (*Aedes*)	75– 750	2–365	6	2– 3	15– 60	2–12
Silkworm	300– 500	9– 12	21– 25	14–21	2– 3	1– 2
Walking stick	200	90	110	–	180	1
Yellow mealworm beetle	276	12– 16	120–150	18–20	60– 90	1

Table 1.9.7: Reproductive and Developmental Data of Selected Butterflies Native to Germany Over the Course of the Year

For the length of individual stages of selected insects, see Table 1.9.6. Monthly data separated by semicolons (;) indicate the appearance of several generations.

Figures from AMANN 1976, FLINDT 1985, HIGGINS and RILEY 1971

	Imago	Egg	Caterpillar	Pupa
Admiral	V–VI; VII–VIII IX–X	IV–VII	VI–X	IX–X
Brimstone	VII–VI	VI	VI–VII	VII
Comma	VI–VII; VIII–V	V; VII VII–VIII	V–VI;	VI; VIII
Eyed hawk moth	VI–VII	VI	VI–IX	IX–VI
Garden tiger	VII–VIII	VII–VIII	VIII–V	VI–VII
Gypsy moth	VII–IX	IX–V	V–VI	VI–VII
Hummingbird hawk moth	VII–X		VI–VIII	VII–VIII
Large yellow underwing	VI–VII; VIII–IX	VIII	IX–V	V–VI
Map butterfly	IV–VI; VII–VIII	V; VII VIII–IX	VI–VII;	VII; IX–V
Mourning cloak	VII–VI	V–VI	VI–VII	VII
Oak processionary	VIII–IX	IX–V	V–VII	VIII
Orange tip	IV–VI	V	V–VII	VIII–IV
Peacock	VII–V	V–VI	V–VII	VII
Pine hawk moth	V–VII	V–VI	VI–X	X–V
Pine moth	VI–VII	VII	VIII–VI	VI–VII
Privet hawk moth	V–VII	VI	VI–IX	IX–VI
Red underwing	VII–IX	VIII–V	V–VI	VI–VII
Silver-washed fritillary	VI–IX	IX–IV	IV–VI	VI
Small tortoiseshell	VI–VII; VIII–VI	V; VII VII–VIII	V–VI; VIII	VI–VII;
Small white (butterfly)	IV–VI; VII–VIII; IX–X		VI–VII; VIII–X	VII–VIII; X–V
Summer fruit tortrix moth	VIII	IV	V–VI o. V–IV	VII
Swallowtail butterfly	IV–VI; VII–VIII	V–VI; VIII	V–VI; VIII–IX	VI–VII; VIII–IV
Winter moth	X–XII	XI–III	III–VI	VI–X

Table 1.9.8: Spawning Data of Selected Fish Native to Germany

Data after LADIGES and VOGT 1965

Species	Months	Species	Months
Barbel	V–VII	Gudgeon	V–VI
Bitterling	IV–VI	Loach	III–IV
(Carp) Bream	V–VI	Pike	II–IV
Brook lamprey	II–V	Rainbow trout	XII–V
Brown trout	I–III	Roach	IV–V
Bullhead	III–V	Rudd	IV–V
Carp	V–VII	Salmon	IX–II
Chub	IV–VI	Sea lamprey	V–VII
Crucian carp	V–VI	Sticklebacks (gasterosteidae)	III–VII
Eurasian minnow	IV–VI	Sturgeon	VI–VII
European perch	III–VI	Wels catfish	V–VII
Grayling	III–VI	Whitefishes	IX–XII

Table 1.9.9: Number and Size of Eggs per Clutch in Selected Fish

The size of the clutch is given as the number of eggs per kilogram body weight (BW). For absolute numbers, see Table 1.9.10. For figures for birds, see Tables 1.9.12 and 1.9.13. Listed according to relative size of clutch.

From BOGNER and RITTER 1976, ZISWILER 1976

Species	Egg diameter (mm)	Clutch size (number of eggs per kg BW)
Turbot	0.9– 1.2	1,200,000
Burbot	1.0– 1.5	1,000,000
Mackerel	0.9– 1.4	800,000
Cod	1.2– 1.8	500,000
Plaice	1.6– 2.2	150,000
European perch	2.0– 2.5	100,000
Carp	1.2– 1.5	100,000
Herring	1.2– 1.5	100,000
European smelt	0.9– 1.0	50,000
Pike	2.5– 3.0	30,000
Sturgeon	2.0– 3.0	25,000
Grayling	3.2– 4.0	8,000
Trout	3.5– 5.5	2,500
Salmon	5.5– 6.0	2,000
Sea or shark catfish	19.0–22.0	50

Table 1.9.10: Number of Annual Offspring (Eggs, Larvae, Young) in Fish, Amphibians, and Reptiles

For egg sizes and clutch sizes relative to body weight, see Table 1.9.9.

Figures from GRZIMEK 1970, *Kosmos-Lexikon* 1953, ZISWILER 1976

Species	Number of progeny
Fish	
Bitterling	50 eggs
Blue shark	30–50 young
Brown trout	up to 1,500 eggs
Carp	up to 700,000 eggs
Cat shark	18–20 eggs
Cod	6,500,000 eggs
Electric rays	10–14 eggs
Herring	30,000 eggs
Manta ray	1 young
Salmon	up to 40,000 eggs
Sandtiger shark	2 eggs
Sticklebacks (gasterosteidae)	60–120 eggs
Sturgeon	6,000,000 eggs
Turbot	9,000,000 eggs
Zander	up to 500,000 eggs
Amphibians	
Alpine salamander	1–2 young
Common frog	1,000–4,000 eggs
Common toad	5,000–8,000 eggs
Edible frog	10,000 eggs
Fire salamander	20–40 larvae
Fire-bellied toads	50–80 eggs
Green toad	10,000–12,000 eggs
Midwife toad	60–80 eggs
Northern crested newt	100–200 eggs
Tree frog	800–1,000 eggs
Reptiles	
African pancake tortoise	1 egg
American alligator	20–70 eggs
Anaconda	30–40 young
Common lizard	3–10 young
Common wall lizard (2–3 clutches)	2–8 eggs
Crossed viper	5–18 young
European pond turtle	3–16 eggs
Galapagos giant tortoise	10–14 eggs
Grass snake	up to 40 eggs

(Cont.) →

Species	Number of progeny	
Hermann's tortoise	5- 8	eggs
Leatherback turtle	80–100	eggs
Nile crocodile	30–50	eggs
Nile monitor	40–60	eggs
Python	100	eggs
Rattlesnake	10- 20	young
Sand lizard	5–14	eggs
Slowworm	5- 15	eggs
Smooth snake	3- 15	young

Table 1.9.11: Reproductive Data of Amphibians and Reptiles Native to Germany
Months are given in which eggs are laid or youth are born. For number of offspring, see Table 1.9.10.

Data after FLINDT 1985, MERTENS 1947

Art	Months
Amphibians	
Agile frog	III–IV
Common frog	II–IV
Common spadefoot toad	III–V
Common toad	II–IV
Edible frog	V–VI
European fire-bellied toad	V
Fire salamander	III–X
Green toad	IV–VI
Midwife toad	V–VI
Moor frog	III–IV
Natterjack toad	IV–VI
Newts	III–V
Tree frog	IV–V
Yellow-bellied toad	IV–IX
Reptiles	
Aesculapian snake	VI–VII
Common lizard	VI–VIII
Common wall lizard	V–VI
Crossed viper	VIII–IX
Dice snake	VI
Grass snake	VII–VIII
Sand lizard	V–VII
Slowworm	VIII–IX
Smooth snake	VIII–IX

Table 1.9.12: Number of Eggs per Clutch in Selected Bird Species

Figures for cuckoo are per breeding period. As a point of comparison: battery farm hens lay on average 265 eggs a year (data from 1985).
For egg size, see Table 1.9.13.

Data after BERNDT and MEISE 1959, HARRISON 1975, HOEHER 1972, MICHAELIS 1974

Species	Clutch size	Species	Clutch size
Albatross	1	Gray heron	3– 5
Barn owl	4– 6	Gray partridge	8–18
Barn swallow	4– 5	Great cormorant	3– 4
Black grouse	6–12	Great crested grebe	4
Black woodpecker	4– 5	Great spotted woodpecker	4– 8
Blackbird	4– 6	Great tit	8–10
Blackcap	5	Greylag goose	5– 6
Black-headed gull	3	Griffon vulture	1
Blue tit	10–13	Herring gull	2– 3
Budgerigar	4– 6	House sparrow	5– 6
Carrion crow	5	Hummingbird (*Mellisuga*)	2
Chaffinch	4– 6	Jackdaw	5
Chiffchaff	5– 6	King penguin	1
Common buzzard	2– 3	Magpie	6– 7
Common raven	4– 6	Mallard	7–11
Common swift	2	Mute swan	5– 8
Cuckoo	15–16	Nightingale	4– 6
Eagle owl	2– 3	Northern lapwing	4
Eurasian coot	5– 9	Ostrich	15
Eurasian jay	5– 6	Pheasant	8–12
Eurasian nuthatch	6– 8	Robin	5– 6
Eurasian reed warbler	4	Skylark	3– 5
European kingfisher	6– 7	Stork	3– 5
European starling	5– 7	Turkey	8–12
Flamingo	1	White wagtail	5– 6
Goldcrest	8–11	Winter wren	6– 7
Golden eagle	2	Wood grouse or Capercaillie	6–10
Golden oriole	3– 4	Wood pigeon	2– 3
Goshawk	3– 4	Yellowhammer	3– 5

Table 1.9.13: Egg Size and Weight in Selected Bird Species

For egg sizes of other groups of animals, see Table 1.9.1. For bird clutch size, see Table 1.9.12. As a point of comparison: a chicken egg weighs on average 50 g.

Data after BERNDT and MEISE 1959, GRZIMEK 1970, HARRISON 1957, HOEHER 1972, STARCK 1975, ZISWILER 1976

Species	Egg size (mm)	Egg weight (g)
Aepyornis (Madagascar's extinct elephant bird)	340 × 250	10,000.0
Barn swallow	20 × 13	2.0
Black redstart	19 × 15	2.0
Blackbird	29 × 22	6.0– 9.0
Black-headed gull	52 × 37	38.0
Budgerigar	19 × 16	2.0
Carrion crow	40 × 30	19.0
Chaffinch	19 × 15	2.1
Chiffchaff	15 × 12	1.4
Common buzzard	56 × 45	60.0
Common raven	50 × 34	30.0
Common swift	25 × 16	3.6
Eagle owl	60 × 50	75.0
Emperor penguin	131 × 86	450.0
Eurasian coot	52 × 36	38.0
European kingfisher	23 × 19	4.5
European starling	30 × 21	6.5
Goldcrest	14 × 10	0.7
Golden eagle	76 × 58	140.0
Gray heron	61 × 43	60.0
Great crested grebe	53 × 36	40.0
Great spotted woodpecker	26 × 19	5.2
Great tit	18 × 13	1.6
Greylag goose	86 × 57	176.0
Griffon vulture	92 × 70	252.0
House sparrow	22 × 16	3.0
Hummingbird (*Mellisuga*)	13 × 8 (!)	0.2 (!)
Jackdaw	35 × 25	12.0
Long-eared owl	40 × 32	23.0
Magpie	34 × 24	10.0
Mallard	58 × 41	50.0
Mute swan	114 × 74	350.0
Nightingale	21 × 16	2.7
Northern lapwing	46 × 33	25.0
Ostrich	150 ×130 (!)	1,500.0 (!)
Peregrine falcon	52 × 41	48.0
Pheasant	45 × 36	30.0
Skylark	23 × 17	3.4

Species	Egg size (mm)	Egg weight (g)
Stork	77 × 53	118.0
Winter wren	16 × 12	1.3
Wood pigeon	40 × 29	19.0

Table 1.9.14: Length of Time of Incubation and Nesting (in Nidicolous Members) of Selected Bird Species

Lengths given in days. *NF* nidifugous species.
As a point of comparison: incubation time for a domestic chicken is 21 days.

Figures from BERNDT and MEISE 1959, HARRISON 1975, HEINROTH 1955, HOEHER 1972, MICHAELIS 1974, PFLUGFELDER 1962, SPECTOR 1956, ZISWILER 1976

Species	Incubation time (days)	Nesting time (days)
Barn swallow	14–16	21
Black redstart	13	13–16
Blackbird	13–14	13–16
Black-headed gull	22–24	NF
Budgerigar	18	30
Canary	12–14	16–20
Carrion crow	18–20	30–32
Chaffinch	12–13	13–14
Chiffchaff	13–14	13–16
Common buzzard	33–35	42–49
Common raven	20–21	40–50
Common swift	18–20	38–54
Cuckoo	12	22–23
Eagle owl	31–37	70
Eurasian coot	21–24	NF
European kingfisher	19–21	26–28
European starling	14–15	21
Golden eagle	43–45	75–80
Great crested grebe	25–29	NF
Great spotted woodpecker	12–13	21
Great tit	13–14	15–18
Griffon vulture	48–50	90
House sparrow	12–14	15–16
Hummingbirds	14–17	19–29
Jackdaw	17–18	30–35
King penguin	52	NF
Long-eared owl	27–28	21–24
Magpie	17–18	25–27

(Cont.) →

Species	Incubation time (days)	Nesting time (days)
Mallard	26–29	NF
Mute swan	34–38	NF
Nightingale	13	11–12
Northern lapwing	24–29	NF
Ostrich	42	NF
Peregrine falcon	28–29	35–38
Pheasant	23–25	NF
Royal albatross	80 (!)	243 (!)
Skylark	11–14	9–10
Stork	31–34	54–55
Winter wren	14–16	15–17
Wood pigeon	15–17	21–28

Table 1.9.15: Incubation Dates of Selected, Well-Known Birds Native to Germany

For incubation and nesting times, see Table 1.9.14.
Roman numerals are used to indicate the months. The Arabic numbers 1, 2, and 3 after the slash stand for the first, second, and third "third" of the month.

Figures from FLINDT 1985

Species	Incubation in months	Species	Incubation in months
Barn swallow	IV–VIII	Eurasian curlew	IV–VI/1
Black redstart	IV–VIII	Eurasian jay	IV–VI
Blackbird	III–VII	Eurasian nuthatch	IV–V
Blackcap	V–VII	Eurasian reed warbler	V–VIII
Black-headed gull	IV/3–VI/1	European goldfinch	V–VI
Blue tit	IV–VI	European greenfinch	IV–VI
Carrion crow	III–VI	European nightjar	V–VI
Chaffinch	IV–VI	European serin	IV–VI
Chiffchaff	IV/3–VI	European starling	IV–VI
Common buzzard	III/3–VII/1	Firecrest	V–VII
Common kestrel	IV–VII/2	Garden warbler	V–VI
Common moorhen	IV/2–V/2	Golden oriole	V–VI
Common redstart	V–VI	Grasshopper warbler	IV/3–VII
Common swift	V–VI	Gray heron	III/3–VI
Crested lark	IV/3–VI	Gray partridge	IV/2–VI/1
Dunnock	III–VII	Great reed warbler	V–VI
Eurasian bullfinch	V–VII	Great spotted woodpecker	V/3–VII
Eurasian coot	IV/3–V	Great tit	IV–VII

Species	Incubation in months	Species	Incubation in months
Green woodpecker	IV/2–VI/2	Rook	III–VI
Hoopoe	V–VI	Sedge warbler	V–VII
House martin	IV–VIII	Skylark	III–VIII
Jackdaw	IV–V	Snipe	IV/3–VII
Lesser whitethroat	V–VII	Song thrush	IV–VIII
Linnet	III–VIII	Sparrow	IV–VIII
Long-eared owl	III/2–V/2	Sparrowhawk	V–VI/1
Magpie	IV–VII	Spotted flycatcher	V–VI
Mallard	III–V	Stork	IV–V
Marsh tit	IV–VI	Tawny owl	III/2–V
Marsh warbler	V/3–VI	Tree pipit	V–VII
Mistle thrush	IV–VII	White wagtail	IV–VII
Mute swan	IV–V	Whitethroat	V–VI
Nightingale	V–VI	Willow warbler	IV/3–VI
Oystercatcher	V–VI/2	Winter wren	V–VII
Pheasant	III/3–VI	Wood pigeon	IV/3–VIII/1
Red kite	IV–VI	Wood warbler	V/2–VI
Red-backed shrike	V–VII	Woodlark	III–VII
Reed bunting	V–VIII	Wryneck	V–VI
Robin	IV–VII	Yellowhammer	IV–VIII

Table 1.9.16: Composition of the Chicken Egg

From STARCK 1975

	Yolk 32 %	Albumen 57 %	Shell 1 %	Total
Water	50 %	86 %	–	65 %
Proteins	16 %	13 %	–	13 %
Fats	32 %	–	–	10 %
Calcium	–	–	94 %	⎫ 11 %
Keratin of the shell	–	–	4 %	⎭
Additional substances	2 %	1 %	2 %	1 %

Table 1.9.17: Embryonal Development of the Chick

Table 1.9.18 gives data on the times at which embryonal stages are reached.

After PFLUGFELDER 1962, modified

Age	Weight (g)	Daily relative weight increase (%)
1 day	0.0002	
2 days	0.003	1,400
3 days	0.02	570
4 days	0.05	150
5 days	0.13	160
6 days	0.29	125
7 days	0.57	97
8 days	1.15	102
9 days	1.53	33
10 days	2.26	48
11 days	3.68	63
12 days	5.07	38
13 days	7.37	45
14 days	9.74	32
15 days	12.0	23
16 days	15.98	33
17 days	18.59	16
18 days	21.83	17
19 days	25.62	17
20 days	30.21	18
21 days (Chick hatches)	Approx. 35	16

Table 1.9.18: Embryogenesis in Selected Mammals and Birds

Times given are from fertilization until the stage in question (h = hours). On embryogenesis, see also Siewing 1980.

Data after ALTMAN and DITTMER 1972, AUSTIN and SHORT 1978, LANGMANN 1976, SPECTOR 1956

Species	2-cell stage (h)	4-cell stage (h)	Start of implantation (days)	Primitive streak (days)	Eyes closed (days)	Eyes open (days)	Birth or hatching (days)
Chicken	3	3.25	–	1.5	13	21	22
Golden or Syrian hamster	16	40	4.5	6.5–7	–	–	16
Human	38	48	6.5	19	70	140	267

Species	2-cell stage (h)	4-cell stage (h)	Start of implantation (days)	Primitive streak (days)	Eyes closed (days)	Eyes open (days)	Birth or hatching (days)
Opossum	40	56	6	6.5	12.25	72	12.5
Pig	30	40	–	11	50	90	112
Rabbit	8	11	7	6.5	19	42	32
Rat	24	48	6	8.5	18	38	22
Rhesus monkey	24	36	9	19	48	–	164
Sheep	30	34	10	13	42	84	150
Sparrow	–	–	–	1.5	11	20	14

Table 1.9.19: Relative Length of Time of Sexual Maturation in Selected Mammals

Sexual maturation time as percentage of average lifespan. For age at sexual maturity, see Table 1.9.4; for lifespan, see Table 1.1.2.
Listed according to percentage, in ascending order.

After KRUMBIEGEL 1955

Species	Maturation time as as share of lifespan (%)	Species	Maturation time as share of lifespan (%)
Polar bear	4	Wild cat	10
Wild boar	6	Marmot	11
Hare	6	Chamois	12
Marmoset	6	Camel	13
Roe deer	7	Brown bear	14
Rat	8	Ermine	15
Fat dormouse	8	Moose (in Europe: elk)	15
Hippopotamus	9	Baboon	16
Fox	10	Human	21
Eurasian badger	10	Chimpanzee	25

Table 1.9.20: Duration of Cycle and Estrus in Selected Mammals

Data after Austin 1967, Austin and Short 1978, Kolb 1974, Niethammer 1979

Species	Cycle duration	Estrus duration
Cat	14 days	7–10 days
Cattle	20 days	12–22 h
Chimpanzee	35 days	–
Common house mouse	4– 5 days	10–20 h
Dog	–	9 days
Fox	–	2–4 days
Golden or Syrian hamster	4 days	12 h
Guinea pig	17 days	6–11 h
Horse	22 days	4– 6 days
Human	28 days	–
Pig	21 days	2– 3 days
Rat	5 days	10–20 h
Rhesus monkey	28 days	–
Sheep	16 days	30–36 h

Table 1.9.21: Duration of Gravidity (Gestation Time) and Number of Young of Selected Mammals

Gestation time is normally the same as the fetal period. In some animals, e.g., in the marten or roe deer, an "extended gestation" results from an ova rest period. In these cases, the actual time of fetal development is given in parentheses. The greatest number of offspring in humans has been recorded for the wife of the Russian farmer F. Wassilet. She gave birth 27 times, bringing a total of 69 children into the world. She had 16 sets of twins, 7 sets of triplets, and 4 sets of quadruplets.

Data after Altman and Dittmer 1972, Austin and Short 1978, Bertelsmann 1979, Harms 1966, Karcher 1975, Krumbiegel 1953, Meyer 1964, Michel 1986, Niethammer 1979, Pflugfelder 1962, Slijper 1956, Starck 1975, Weismann 1976, Wood 1982, Ziswiler 1976

Species	Duration of gravidity (days)	Offspring per litter or birth (n)
Bat	50	1
Beaver	180	2–4
Beech marten	260 (42)	3– 5
Blue whale	320–360	1
Brown bear	240	2–4
Camel	390–398	1
Cat	55– 66	4
Cattle	280	1
Chamois	147–190	1– 2
Chimpanzee	231–253	1

Species	Duration of gravidity (days)	Offspring per litter or birth (n)
Common house mouse	21– 24	3–10
Common or Harbor seal	300	1
Common shrew	20	6– 7
Common vole	21	4
Dog (according to race)	59– 65	1–10
Dolphin	276	1
Donkey	365	1
Dromedary	388	1
Elephant	623–660 (!)	1
Eurasian badger	200 (56)	2–5
Fat dormouse	28	3– 7
Fox	49– 55	3– 7
Gibbon	210	1
Giraffe	431–465	1
Goat	150	2– 3
Golden or Syrian hamster	16	1–12
Gorilla	251–289	1
Guinea pig	63– 68	2
Hare	42	2– 4
Hedgehog	34– 49	4–10
Hippopotamus	237–247	1
Horse	314–373	1
Human	274	1
Kangaroos	30– 40	1
(+ in pouch)	135–240	
Lion	108	2– 6
Macaque	168	1
Mole	28– 42	1– 6
Opossum	13 (!)	up to 12
Orangutan	233	1
Pig	114	6–12
Polar bear	240	2– 4
Polecat	180 (43)	3–7
Rabbit	30– 32	4–12
Rat	21	7– 9
Red deer (Wapiti)	234	1
Red kangaroo	38– 40	1
Reindeer	242	1
Rhesus monkey	163	1
Rhinoceros	548–578	1
Roe deer	280 (150)	1– 2
Squirrel	35	3
Tiger	105	2– 3
Walrus	330–370	1

(Cont.) →

Species	Duration of gravidity (days)	Offspring per litter or birth (*n*)
Wild boar	133–139	4–10
Wolf	62– 66	4– 9
Wood mouse	25	2– 3

Table 1.9.22: Relative Birth Weight of Selected Mammals

Weight given as percentage of weight of mother. Listed according to relative size, in ascending order. Here, see also Table 1.1.10.

Figures from KRUMBIEGEL 1955, NIETHAMMER 1979, SCHULTZ 1972, SLIJPER 1967, ZISWILER 1976

Species	Relative birth weight (%)	Species	Relative birth weight (%)
Kangaroo	0.0003	Human	5.0
Siberian brown bear	0.166	Baboon	7.1
Puma	0.7	Spiny mouse	9.0
Rat	1.0	Common or Harbor seal	10.0
Hippopotamus	1.0	Etruscan shrew	10.0
Chimpanzee	2.5	Marmoset	12.1
Gorilla	2.6	Brazilian porcupine	30.0
Blue whale	3.0	Dolphin	30.0
African elephant	4.0	Bats	10–40.0

Table 1.9.23: Normal Growth (Mean Values of Body Weight) of Selected Pets and Laboratory Animals

Numbers apply to the most frequently used animal lines in research. Figures in grams. For maximum weights, see Table 1.1.10.

Data after ALTMAN and DITTMER 1972

	Guinea pigs ♂	♀	Golden or Syrian hamster ♂	♀	Common house mouse ♂	♀	Brown rat ♂	♀
1 day	80	78	2.9	3.0	1.9	1.9	6.7	6.5
1 week	115	110	6.9	7.8	4.4	4.3	18.2	16.0

	Guinea pigs		Golden or Syrian hamster		Common house mouse		Brown rat	
	♂	♀	♂	♀	♂	♀	♂	♀
2 weeks	155	145	18.3	17.5	6.5	6.0	43.0	40.8
3 weeks	200	180	40.0	40.3	10.7	9.6	59.4	55.6
4 weeks	240	218	48.7	44.2	13.7	12.4	98.9	94.9
6 weeks	322	297	86.1	93.0	21.7	20.9	162.3	153.4
8 weeks	405	374	91.5	94.5	25.5	22.8	235.7	202.0
10 weeks	482	449	99.5	103.3	28.5	25.8	292.7	239.1
12 weeks	546	518	103.9	114.9	33.4	29.9	358.7	246.6
16 weeks	656	625	121.9	135.9	34.5	32.4	388.0	259.9
20 weeks	760	720	131.8	149.6	36.7	34.7	439.4	281.2
24 weeks	840	796	140.5	157.8	39.8	36.8	452.0	288.4
28 weeks	899	837	–	–	–	–	491.2	291.8

Table 1.9.24: Times for Doubling Birth Weight in Selected Mammals

Number of days required until birth weight is doubled. For the development of weight in selected species, see Table 1.9.23. Listed in order of size, from shortest to longest times required.

Figures from OPPENHEIMER and PINCUSSEN 1927

Species	Time needed for doubling weight (days)
Mouse	4 – 6
Rat	5 – 9
Rabbit	6
Pig	6.5– 16
Dog	8
Cat	9.5
Sheep	10 – 12
Guinea pig	13 – 17
Goat	18 – 20
Common or Harbor seal	42
Cattle	47
Horse	60
Donkey	70
Human	127 –180

Table 1.9.25: Composition of Milk in Selected Mammals

Data after *Documenta Geigy* 1975, GRZIMEK 1970, HIRSCH 1977, KLEIBER 1967, Meyer 1964, OPPENHEIMER and PINCUSSEN 1925, SPECTOR 1956

Species	Water (%)	Protein (%)	Fat (%)	Carbohydrates (%)
Baboon	85.4	1.6	5.1	7.5
Black bear	79.5	7.6	10.9	1.6
Blue whale	45.0	12.0	42.0	1.0
Buffalo	89.6	4.0	1.8	4.6
Camel	87.1	3.7	4.2	4.1
Cat	82.3	9.1	3.3	4.9
Cattle	88.5	3.2	3.7	4.6
Dog	81.5	6.8	8.6	4.1
Dolphin	51.0	4.0	44.0	1.0
Donkey	90.0	2.0	2.0	6.0
Elephant	66.7	3.2	22.1	7.4
Fox	81.6	6.6	5.9	4.9
Giraffe	78.3	5.8	12.5	3.4
Goat	86.0	3.6	4.2	4.8
Guinea pig	81.9	7.4	7.2	2.7
Horse	91.1	2.1	1.25	6.3
Human	87.7	1.03	4.4	6.9
Mouse	60.0	12.0–13.0	21.0–33.0	2.0
Orangutan	88.5	1.4	3.5	2.7
Pig	80.9	5.8	8.5	4.8
Rabbit	71.0	10.5	16.5	2.0
Rat	68.0	12.0	15.0	3.0
Red kangaroo	78.0	6.0–8.0	2.0–10.0	4.0–15.0
Reindeer	64.7	10.3	22.5	2.5
Rhinoceros	89.3	3.2	0.3	7.2
Roe deer	76.0	9.0	10.5	4.5
Sheep	81.6	5.6	7.5	4.4
Yak	72.5	5.2	7.0	5.3

Table 1.9.26: Annual Milk Production of Domestic Animals

Figures represent averages. Cattle reach top production values of over 10,000 liters annually. High-performance cows in Israel produce up to 13,000 l milk per year.

Figures after BERTELSMANN 1979, MANN 1979

Species	Average annual milk production (liters)
Cattle	4,000–5,000
Goat	600– 800
Sheep	500– 700
Zebu	180– 250

Table 1.9.27: Lactation Periods of Selected Mammals

Figures from FRÄDRICH and FRÄDRICH 1973, KOLB 1974

Species	Lactation period	
Brown bear	3– 4	months
Cat	4– 6	weeks
Cattle	8–10	weeks
Chimpanzee	> 1	year
Common or Harbor seal	4– 6	weeks
Deer	3– 4	months
Dog	6	weeks
Donkey	12–20	weeks
Elephants	up to 2	years
Fox	7– 8	weeks
Gorilla	1.5	years
Guinea pig	2– 3	weeks
Hedgehog	18–20	days
Horse	12–20	weeks
Lion	10	weeks
Orangutan	3– 4	years
Pig	8–10	weeks
Rabbit	8	weeks
Rhesus monkey	7–14	months
Rhinoceri	2	years
Roe deer	2– 3	months
Sheep	6–12	weeks
Tiger	8	weeks
Wild boar	2– 3	months
Wolf	7– 8	weeks

Table 1.9.28: Closed Seasons for Huntable Animals in Germany

Figures from *DJV-Handbuch* 1978

	Closed season (months)
Red deer (in N. Am.: elk), ♂, ♀	II–VII
Red deer (in N. Am.: elk), calves	III–VII
Fallow deer, ♂, ♀	II–VIII
Fallow deer, calves	III–VIII
Roe deer, bucks	15. X–15. V
Roe deer, does	II–VIII
Chamoix	15. XII–VII
Wild boars	II–15. VI
Hares	15. I–IX
Beech and pine martens	III–15. X
Badgers	IX–VII
Capercaillie	VI–IV
Partridges	15. XII–VIII
Pheasants	15. I–IX
Pigeons (Wood pigeons, Eurasian collared doves)	V–VI
Mallards	15. I–VIII
All other wild ducks	15. I–IX
Wild geese	15. I–X
Eurasian woodcocks	15. I–15. X

1.10 Nervous System

Table 1.10.0: Numbers that Amaze

This table brings together some interesting numbers from the Sects. 1.10 Nervous System and 1.11 The Senses, which are cited again in the individual tables below, where they are accompanied by references to the literature.

Absolute brain weight		
	in the blue whale	4,925 g
	in the elephant	4,700 g
Relative brain weight as a percentage of body weight	in the spider monkey	6.6 %
	in the sparrow	2.9 %
As a point of comparison	in the human	2.5 %
Conduction velocity in Aα fibers in mammals		120 m/s = 432 km/h
The electrical discharge		
	in the electric eel	up to 700 V
	in the electric ray	up to 350 V
Number of photoreceptors in the eye		
	in the brown rat	1,400,000
As a point of comparison	in the human	160,000
The visual acuity angle (minimum separable angle under which two points can still be optically distinguished)	in the peregrine falcon	25"
The number of ommatidia (unit eyes) per compound eye in arthropods		
	in the bee (drone)	7,500
	in the dragonfly	28,000
	in the ghost or sand crab	30,000
Flicker fusion rate of the eye (rate of images below which individual images are seen)	in dragonflies	200
	in pigeons/doves	148
As a point of comparison	in the human	21
Upper limit of hearing	in bats	400 kHz
	in nocturnal butterflies	240 kHz
	in the dolphin	200 kHz
As a point of comparison	in a young human	21 kHz

Table 1.10.1: Brain Weight (Absolute) of Selected Mammals

Figures represent mean values. Since absolute values, due to great variations in size within individual species, have little explanatory power, see the relative brain weights given in Table 1.10.2. For humans, see also Table 4.6.2.

Figures after HALTENORTH 1977, KRUMBIEGEL 1953

Species	Brain weight		Species	Brain weight	
Blue whale	4,700	g	Hedgehog	3.4	g
Cat	31.4	g	Hippopotamus	580	g
Cattle	540	g	Horse	590	g
Chimpanzee	350–400	g	Human	1,200–1,500	g
Dog	135	g	Lion	222	g
Dolphin	700	g	Mouse	0.4	g
Elephant	4,925	g	Opossum	6.5	g
Gibbon	118	g	Orangutan	370– 400	g
Gorilla	430– 650	g	Pig	112	g
Guenon	106	g	Rat	3.05	g
Harbor porpoise	455	g	Wolf	114.4	g

Table 1.10.2: Relative Brain Weight of Selected Vertebrates

Figures given as percentage of body weight. A comparison only makes sense within closely related animal forms. For the absolute brain weights of selected species, see Table 1.10.1, and for the cerebralization index, see Table 1.10.4.

Data after BERTELSMANN 1979, HALTENORTH 1977, *Kosmos-Lexikon* 1953, KRUMBIEGEL 1953

Species	Relative brain weight (%)	Species	Relative brain weight (%)
Blue whale	0.007	Horse	0.25
Carp	0.4	Human	2–2.5
Cat	0.94	Lion	0.2
Cattle	0.07	Mouse	3.2
Chimpanzee	0.8	Orangutan	0.66
Dog	0.59	Ostrich	0.08
Eagle	0.6	Rat	1.22
Elephant	0.08	Shark	0.04
Frog	0.6	Sheep	0.33
Golden or Syrian hamster	0.88	Sparrow	2.9
Gorilla	0.2	Spider monkey (*Ateles*)	6.6 (!)
Guinea pig	1.33	Tuna	0.003
Hippopotamus	0.05	Wolf	0.52

Table 1.10.3: Brain Surface Area of Selected Mammals

Figures refer to the surface area of one hemisphere in square millimeters.

After SALLER 1964

Species	Surface area of one hemisphere (mm^2)
Cattle	49,849
Chimpanzee	39,572
Elephant	301,843
Gibbon	16,301
Goat	12,005
Guenon	14,641
Harbor porpoise	46,994
Horse	56,995
Human (European, of average size)	112,471
Pig	13,022
Sheep	14,014

Table 1.10.4: Cerebralization Index of Birds and Mammals

The cerebralization index or ratio is the quotient of the weight of the cerebrum spheres (in birds) or of the neopallium (in mammals) divided by the weight of the brain parts considered primal.
Listed in ascending order for birds and mammals, respectively.

After PORTMANN 1959 and 1965, STARCK 1974

Species	Cerebralization index	Species	Cerebralization index
Pheasant	3.18	Rabbit	4.6
Loon	3.69	Squirrel	6.15
Herring gull	4.31	Polecat	12.9
Little egret	5.32	Prosimian (mongoose lemur)	13.5
Mallard	6.08	Wild boar	14.1
Goshawk	7.24	Fox	16.8
Common raven	18.95	Zebra	29.5
Parrot (red-and-green macaw)	27.61	Guenon	33.9
		Baboon	47.9
Eurasian water shrew	0.759	Ape	49
Hedgehog	0.776	Asian elephant	104
Bat	1.13	Atlantic white-side dolphin	121
Muskrat	2.75	Human	170

Table 1.10.5: Distribution of Important Ions on the Cell Membranes of Selected Excitable Structures

Figures in millimoles. For data on resting potentials, see Table 1.10.6.

Figures from GAUER et al. 1971, PENZLIN 1970, SCHMIDT 1974

	Potassium Inner	Outer	Sodium Inner	Outer	Chloride Inner	Outer
Cuttlefish, giant nerve fibers	410	10	50	460	60	540
Frog, nerve	120	2.5	37	120		
Frog, skeletal muscle	130	2.5	15	120	10	120
Rat, skeletal muscle	152	4	16	150		
Rat, cardiac muscle	140	2.7	13	150		
Cat, skeletal muscle	150	4	12	150	4	140
Cat, motoneuron	150	5.5	15	150	9	125

Table 1.10.6: Resting Potentials of Nerve and Muscle Fibers

On the generation of resting potential, see SIEWING 1980.

Figures from FLOREY 1970, LAGET 1970, PENZLIN 1970, SCHMIDT 1974

Muscle fibers

Frog	−82 up to −100 mV
Rat	−100 mV
Dog, ventricle	−65 up to − 95 mV
Dog, atrium	− 85 mV
Dog, Purkinje fibers	− 90 mV

Nerve fibers

Cuttlefish (Sepia), in situ	− 70 mV
Cuttlefish (Sepia), ex situ	− 62 mV
Loligo	− 60 mV
Earthworm	− 70 mV
Cambarus	− 90 mV
Cockroach	− 70 mV
Carcinus	−70 up to − 94 mV
Frog, myelinated fiber	− 71 mV

Table 1.10.7: Nerve Conduction Velocity

For the corresponding figures for plants, see Table 2.6.18.

After FLOREY 1970, GAUER et al. 1971, HANKE et al. 1977, KATZ 1971, LAGET 1970, SCHMIDT 1974

Animal species	Type of nerve fiber	Diameter (µm)	Nerve conduction velocity (m/s)
Plumose anemone (*Metridium*)	Nerve net		0.13
Moon jelly (*Aurelia*)	Nerve net	6 – 12	0.5
Earthworm	Med. giant fiber	50 – 90	30
	Lat. giant fiber	40 – 60	11.3
Sepia	Giant fiber	200	7
Loligo	Giant fiber	400 (!)	20
Lobster	Leg nerve	60 – 80	14 – 18
Crab	Giant fiber	100 –250	15 – 20
Cockroach	Giant fiber	50	1.5– 3.5
Carp	Lateral fiber	20	47
Frog	A fiber	15	30
Cat	A fiber	13 – 17	78 – 102
	Spinal nerve	5.5	50
	C fiber	0.5– 1	0.6– 2
Human	Aα fiber	10 – 20	60 –120 (!)
	Aβ fiber	7 – 15	40 – 90
	Aγ fiber	4 – 8	30 – 45
	Aδ fiber	2.5– 5	5 – 25
	B fiber	3	3 – 15
	C fiber	0.3– 1.3	0.6– 2.3

Table 1.10.8: Size of the Discharge of Electric Fish

On the anatomy and function of electric organs, see SIEWING 1980.

Figures from FLOREY 1970

Species	Voltage
Electric eel (*Electrophorus electricus*)	700 V
Electric catfish (*Malapterurus electricus*)	350 V
Electric rays (*Torpedo nobiliana*)	60 V
Ray (*Narcine brasiliensis*)	30 V
Ray (*Raja clavata*)	4 V
Elephant fishes (*Mormyridae*)	1– 12 V

1.11 Senses

Table 1.11.1: Minimum Stimuli for Activating a Sensory Organ

Data always for optimal stimuli quality.

Data after AUTRUM 1948

Sensory organ	Minimum stimuli strength
Subgenual organ, cockroach	4×10^{-10} Watt
Ear, grasshopper	5×10^{-17} Watt
Eye, human	5.6×10^{-17} Watt
Ear, human	8×10^{-18} up to 4×10^{-17} Watt

Table 1.11.2: Size of the Eyeball in Selected Animals

Data after KÄSTNER 1973, OPPENHEIMER and PINCUSSEN 1925

Species	Diameter (mm)	Volume (mm³)
Bat	1	
Cat	21	4.9
Cattle	40 – 41	31.8
Chicken		1.78
Common house mouse	3	
Dog	20 – 21	5.1
Elephant	40	
Giant squid	400	
Horse	50 – 51	45.0
Human	24	6.5
Lion	35	
Pig	25 – 26	9.5
Rabbit	16.5	3.0
Rat	6	
Sheep	30.5 – 31	11.65
Tawny owl		8.32

Table 1.11.3: Focal Distance of the Lenses in the Eyes of Selected Animals

The focal distance of a lens (in meters) can serve as the basis for deriving refractive power (in diopters) as its reciprocal value. Thus, the smaller the focal distance of a lens, the greater its refractive power. Focal distance varies according to the degree of accommodation (adjustment to proximity/distance). For this reason, the range is given for humans from non-accommodation to maximum accommodation.

Data after BERNDT and MEISE 1959, BUDDENBROCK 1924, VELHAGEN 1972

Species	Focal distance (mm)
Birds (most)	50
Chickens. pigeons/doves	80 –125
Divers	20 – 25
Owls	250 –500
Blue whale	40
Bowhead whale	20
Bristleworm (*Alciope*)	0.3
Electric rays	2.23
Frog	5
Hare	10
Human	69.9– 40.4
Lion	25.5
Pig	13.3

Table 1.11.4: Number of Photoreceptors in the Eyes of Selected Animals

Figures given per square millimeter. For the number of photoreceptors per axon, see Table 1.11.5; for the number of ommatidia in the compound eyes of arthropods, see Table 1.11.8.

Data after BUDDENBROCK 1924, KÄSTNER 1973, ZISWILER 1976

Species	Photoreceptors (n)
Brown rat	1,400,000
Carp	510,000
Cat	397,000
Cattle	358,000
Chicken	265,000
Cuttlefish (squid)	115,000
Frog	95,000
Human	160,000
Tawny owl	680,000
Trout	303,000

Table 1.11.5: Number of Photoreceptors per Dendrite in the Eyes of Selected Vertebrates

Figures from BUDDENBROCK 1924, SIEBECK 1977, ZISWILER 1976

Species	Photoreceptors/dendrite (n)
Blue whale	5,100
Chameleon, fovea	1–2
Human, fovea	1
Human, periphery	130
Pinniped	1,500
Porpoise	4,900
Shark	40
Shark, core	8
Sparrow	1
Tiger	2,500

Table 1.11.6: Visual Acuity Angle (Angle of Minimum Separability) of Selected Animals

The visual acuity angle designates the greatest angle at which two closely adjacent points can still just be discriminated from each other.
° = degree, ′ = arcminutes, ″ = arcseconds.
As a point of reference: a 17.5-mm-high object 1 m from the eye appears at a 1° angle; a 0.3-mm-high object appears at 1 arcminute (1′).

Figures from BERNDT and MEISE 1959, BUDDENBROCK 1956, NIETHAMMER 1979, PENZLIN 1970, RAHMANN 1967, SIEBECK 1977

Species	Visual acuity angle
Alligator	11′
Asian elephant	10′ 20″
Bat	3–6°
Bee	1°
Blackbird	1′ 20″
Cat	5′ 30″
Chaffinch	1′ 20″
Chicken	4′ 14″
Chimpanzee	28″
Common frog	6′ 53″
Common hermit crab	4° 12′
Donkey	8′ 36″
Eurasian minnow	10′ 50″
European pond turtle	2′ 51″
Fruit fly	9° 17′

Species	Visual acuity angle		
Golden or Syrian hamster	1°	4′	
Horse		3′	15″
Human			25″
Peregrine falcon			25″
Rat (albino)		40′	
Rat (wild)		20′	
Red deer (Wapiti)		11′	18″
Rhesus monkey			34″
Robin		2′	38″
Rock dove		2′	42″
Sand lizard		11′	28″
Shrimp	4°	35′	
Skylark		1′	20″
Yellowhammer		3′	7″

Table 1.11.7: Visual Field (Horizontal) of Both Eyes in Selected Animals

Figures represent mean values. Among primates there is considerable variation, depending on the placement of the eyes.

Data after SIEBECK 1977

Species	Visual field
Barn owl	160°
Common kestrel	300°
Crocodiles	295°
Dog	250°
Frogs	334–360°
Lizards	280–300°
Plaice	360°
Primates	180°
Rock dove	300°

Table 1.11.8: Number of Unit Eyes (Ommatidia) in the Compound Eyes of Selected Arthropods

Presented in ascending order.
For greater detail on anatomy and function, see SIEWING 1980.

After KÄSTNER 1973, OPPENHEIMER and PINCUSSEN 1936, WEBER 1966

Species	Ommatidia (n)
Water flea	22
Woodlouse	30
Yellow meadow ants	80
Earwig	270
Firefly ♀	300
Red wood ant	600
Firefly ♂	2,500
Housefly	3,200
Honeybee (queen)	3,900
Blowfly	5,400
Cockchafer (May bug)	5,500
Shore crab	7,000
Honeybee (drone)	7,500
Walker	12,200
Dragonfly	28,000
Ghost or Sand crab	30,000

Table 1.11.9: Flicker Fusion Rate of the Eyes of Selected Animals

Figures represent images per second. At lower image frequencies, individual images are perceived. Listed in ascending order.

Data after *Lexikon der Biologie*, SIEBECK 1977

Species	Image frequency (per second)
Salamander	5
Human	20
Gecko	20
Beetle	20– 30
Cat	27– 30
Guinea pig	30
Frog	48
Bee	55
Rock dove	148
Dragonfly	300

Table 1.11.10: Absorption Maxima of the Visual Pigments of Selected Invertebrates

Figures represent the wavelength in nanometers at which light is maximally absorbed. For rhodopsin values for vertebrates, see Table 1.11.11.

Figures from ALTMAN and DITTMER 1973, PENZLIN 1970

Species	Maximum absorption at (nm)
Cockroach	500
Cuttlefish (*Sepia*)	492
Honeybee	440
Horseshoe crab	520
Housefly	510, 437
Lobster	515
Octopus (*Octopus*)	475
Shrimp	555, 496
Squid (*Loligo*)	493

Table 1.11.11: Absorption Maxima of the Rhodopsins of Selected Vertebrates

Figures represent the wavelength in nanometers at which light is maximally absorbed. For data on visual pigments in selected invertebrates, see Table 1.11.10. On the basic processes of sight, see SIEWING.

Figures from ALTMAN and DITTMER 1973, PENZLIN 1980.

Species	Maximum absorption (nm) at	
	Retinal 1	Retinal 2
Alligator	499	
Carp		523
Chicken	560, 500	
Common house mouse	498	
Duck	502	
Eel	487	523
Frog	502	
Goldfish		522
Guinea pig	497	
Human	497, 493	
Lamprey	497	518
Monkey	500	
Pelican	502	
Rabbit	502	
Rat	498	
Rock dove	544, 502	
Tadpole	512	
Trout	503	527

Table 1.11.12: Eardrum Surface Area of Selected Animals

On hearing ability, see Table 1.11.13; for humans, see also Table 4.6.10.
Presented in ascending order.

Figures from MANN 1972, REIN and SCHNEIDER 1971

Species	Eardrum surface (mm^2)
Mouse	2.7
Tree frog	6.2
Natterjack toad	7.1
Hedgehog	16.0
Common frog	20.0
Marsh or Lake frog	28.0
Rhesus monkey	32.0
Cat	42.0
Dog	46.0
Human	55.0
Horse	75.0
Bullfrog	78.0
Cattle	86.0
Gorilla	97.0

Table 1.11.13: Upper Limit of Hearing Ability in Selected Animals

For humans, see also Table 4.6.10.

Data after BERNDT and MEISE 1959, BERTELSMANN 1979, GRZIMEK 1970, KÄSTNER 1973, MEYER 1964, NIETHAMMER 1979, PENZLIN 1970, STURKIE 1976

Species	Upper limit (kHz)	Species	Upper limit (kHz)
Axolotl	0.2	Dog	135
Bat	400	Dolphin	200
Brown bullhead	13	Duck	< 8
Budgerigar	14	Eagle owl	8
Bullfrog	4	Eel	0.6
Canary	10	Eurasian bullfinch	21
Cat	47	Eurasian minnow	7
Chaffinch	29	European starling	15
Chicken	38	Freshwater turtle	4
Chimpanzee	33	Goldfish	3
Cicadas	20	Grasshoppers	12
Crickets	8	Guinea pig	33
Crow	< 8	Human	21

Species	Upper limit (kHz)	Species	Upper limit (kHz)
Katydids (Bush crickets)	90	Robin	21
Lizards	8	Rock dove	12
Long-eared owl	18	Sharks	2
Mouse	100	Snakes	0
Nocturnal butterflies	240	Sparrow	18
Penguin	15	Tawny owl	21
Rat	80	Yacare caiman	6

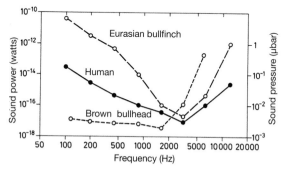

Fig. 1.11.13: Dependency of auditory threshold on frequency in humans and Eurasian bullfinches (auditory threshold data in watts) and in brown bullhead (threshold data in microbars). Cf. also Table 1.11.13, and for humans, Fig. 4.6.11.
From Penzlin 1980, p. 381

Table 1.11.14: Auditory Acuity of Selected Animals

Figures given represent the smallest angle at which two sources of sound can still be discriminated.
As a point of comparison: At a 1° angle at a distance of 10 m, two sources of sound are 17.5 cm apart.

From Buddenbrock 1956

Species	Angle of auditory acuity
Cat	1.5°
Chicken	4°
Dog	2.5°
Eurasian bullfinch	20–25°
Human	8.4°

Table 1.11.15: Number of Taste Buds of Selected Vertebrates

From STURKIE 1976, modified

Species	Taste buds (n)
Bat	800
Cat	473
Cat shark	100,000
Cattle	25,000
Chicken	24
Duck	200
Eurasian bullfinch	46
European starling	200
Goat	15,000
Human	9,000
Parrot	350
Pig	15,000
Rabbit	17,000
Rock dove	37
Snakes	0

Table 1.11.16: Dimensions of Olfactory Epithelium in Selected Animals

Data after STARCK 1975

Species	Surface area (cm^2)	Receptors (n)
Cat	20.8	
Dog	85	2.3×10^8
Human	2.5–5	3×10^7
Rabbit	9.3	1×10^8
Roe deer	90	3×10^8

Table 1.11.17: Vibration Threshold at Optimal Frequency

The vibration threshold designates the amplitude at which a vibration is first perceived.
Since this varies according to frequency, the optimal frequency is also provided.
Presented in ascending order.

Figures from HEIDERMANNS 1957, KÄSTNER 1972

Species	Optimal frequency	Vibration threshold (nm)
Cockroach	1.4	0.004
Wart-biter bush cricket (*Decticus*)	2	0.0366
Field cricket	1.5	0.1
Cockchafer (May bug)	1.8	0.1
Peacock	2	0.68
Bee	2.5	13
Earwig	1.2	19
Carrion beetle	1	40.5
Birds	0.8	100
Human	0.2	1,000
Assassin bugs (*Reduvius*)	0.25	1,000
Flower fly	0.2	6,590
Blowfly (*Calliphora*)	0.3	17,400

Part 2: Botany

For pragmatic reasons, fungi will be left in Part 2, since both the general public and biology as it is taught at school continue to consider them as part of the plant kingdom. [In the original German edition of this volume, the German names of local plants were adopted from SCHMEIL and FITSCHEN 1968.]

For fungi and bacteria, see also Part 4: Microbiology.

2.1 General Data: Dimensions, Age, etc.

Table 2.1.0: Numbers that Amaze

This table brings together several interesting figures from Part 2, which are cited again in the individual tables below, where they are accompanied by references to the literature.

Highest known age in a tree		
	Redwood (sequoia)	4,000 years
	Bristlecone pine	4,600 years
Root depth	Grape vine	12–16 m
	Desert plants (max.)	20 m
Speed of growth		
	Bamboo shoots	57 cm/day
	Banana leaf sheaths	160 cm/day
Single leaf with the greatest surface area		
	Victoria regia (max.)	2 m^2
Beech tree has an overall leaf surface area of up to		446 m^2
Water vessel diameter in	Lianas (max.)	700 µm
This allows a water conductance speed of up to		150 m/h
Heaviest wood	Guaiacum (or lignum vitae)	1.23 g/ cm^3
Lightest wood	Balsa wood	0.18 g/cm^3
Tallest tree	Australian mountain ash (*Eucalyptus regnans*)	152 m
Tree with the greatest trunk diameter	Baobab	15 m
Tree with the greatest trunk circumference	Baobab	47 m
Longest blooming period of a single blossom		
	Odontoglossum rossii	80 days
Blossom diameter	*Rafflesia arnoldii*	1 m
Number of seeds per tree	Birch	30,000,000
Number of seeds per capsule	Swan orchid	3,751,000
Seed weight	Broomrape	0.000001 g
	Avocado	60 g
Ballistic fruits catapult their seeds	Liana *Bauhinia purpurea*	15 m
	Lupin	7 m
Osmotic pressure in cell liquid	Common glasswort	113 atm
	Sod (*Suaeda*)	130 atm

Table 2.1.1: Number of Species of Plant Taxa and Fungi

A total of over 400,000 species have now been described. For more specific figures on seed-bearing plants, see Table 2.1.2.

Data after HEYWOOD 1978, KRAUS and KUBITZKI 1982, STRASBURGER 1983

Prokaryotes (bacteria, cyanobacteria)	3600
Eukaryotic algae	33,000
of which: Green algae	7,000
Diatoms	6,000
Red algae	4,000
Brown algae	1,500
Fungi, including slime molds	90,000
of which: Ascomycota (sac fungi)	30,000
Basidiomycota	30,000
Mosses	26,000
Lichens	20,000
Fernlike plants	15,000
of which: Lycopodiophyta	400
Horsetails	32
Seed-bearing plants (spermatophytes)	236,000
of which: Gymnosperms	800
Angiosperms	235,000

Table 2.1.2: Number of Species Within the Individual Taxa of Seed-Bearing Plants

Figures from HEYWOOD 1978

I: Dicotyledoneae (= Magnoliatae)		177,000
1. Magnoliidae		13,200
Including:	Ranunculales	2,800
	Piperales	3,000
2. Hamamelididae		1,396
Including:	Fagales	1,200
3. Caryophylliidae		11,000
Including:	Caryophyllales	9,700
	Polygonales	750
4. Dilleniidae		29,600
Including:	Malvales	2,700
	Urticales	6,000
	Violales	5,250
	Ericales	3,570
	Primulales	2,000

5. Rosidae		57,700
Including:	Rosales	7,100
	Fabales	17,000
	Euphorbiales	5,100
	Geraniales	2,900
	Umbellales	3,700
6. Asteridae		64,000
Including:	Gentianales	5,600
	Scrophulariales	8,800
	Campanulales	2,200
	Asterales	25,000
II. Monocotyledoneae (=Liliatae)		52,800
1. Alsimatidae		562
2. Commelinidae		19,700
Including:	Poales	9,000
	Cyperales	4,000
	Bromeliales	2,000
3. Arecidae		5,700
4. Liliidae		26,800
Including:	Liliales	8,700
	Orchidales	18,100

Table 2.1.3: Maximum Age of Selected Plants

In forestry, trees are felled before reaching their maximum age. This so-called rotation period is given in Table 2.1.4.

Plants are listed in ascending order, according to age.

After BERTELSMANN 1979, CZIHAK et al. 1976, MEYER 1964, SPECTOR 1956

Species	Age (years)	Species	Age (years)
Lycopodiophyta	7	European hornbeam	150
Oak fern	7	Goat willow	150
Moss (polytrichum moss)	10	Common aspen	150
Blueberry (*Vaccinium*)	25	Apple	200
Common moonwort (fern)	30	Sycamore maple	200
Heather	42	Pear	300
Cyclamen	60	Cherry	400
European rowan	80	Dog rose	400
Mountain avens	100	Walnut	400
Hazelnut	120	Ivy	440
Birch	120	Juniper	500
Grape vine	130	Pine	500

(Cont.) →

Species	Age (years)	Species	Age (years)
Fir	500	Cedar	1,300
Elm	500	Platanus	1,300
Larch	600	Oak	1,300
Norway maple	600	Yew	1,800
Olive	700	Tilia (linden, lime)	1,900
Sweet chestnut	700	Fig (*Ficus religiosa*)	2,000
European beech	900	Cypress (*Cupressus*)	2,000
Spruce	1,100	Redwood (sequoia)	4,000
Swiss pine	1,200	Bristlecone pine	4,600

Table 2.1.4: Rotation Periods of Important Harvested Trees

The average age of trees in forestry is called their rotation period.
Cf. this to the maximum ages of plants given in Table 2.1.3.

From *Schutzgemeinschaft Deutscher Wald* 1967, modified

Species	Rotation period (years)
Ash	100–140
Birch	60– 80
Black alder	60– 80
Black poplar	30– 50
Douglas fir	60–100
Elm	120–140
European hornbeam	60–100
Fir	90–130
Larch	100–140
Norway maple	120–140
Oak	180–300
Pine	100–120
Spruce	80–120
Sycamore maple	120–140

Table 2.1.5: Lifespan of Leaves and Needles of Selected Evergreen Plants

In ascending order, according to age.

Figures from BERTELSMANN 1979, MEYER 1964, MOLISCH 1965

Species	Lifespan (years)	Species	Lifespan (years)
Cherry laurel	1–3	Swiss pine	3–6
Ivy	1–3	Juniper	4–5
Oleander	1–3	Bay laurel	4–6
Privet	1–3	Spruce	4–6
Cowberry (lingonberry)	2	Fir	5–7
Olive	2	Yew	6–8
Pine	2–3	Mountain pine	7.5
English holly	2–4	Norfolk Island pine	15
Redwood (sequoia)	2–4	Welwitschia	100

Table 2.1.6: Root Depth of Selected Plants

Presented in ascending order.
After BERTELSMANN 1979, HAGER 1960, MEYER 1964

Species	Root depth (m)	Species	Root depth (m)
Dandelion	0.30	Pea	2.10
Hellebore	0.35	White lupine	2.10
Cypress spurge	0.50	Red clover	2.10
Swallow-wort	0.60	Barley	2.60
Field bindweed	1.00	Oats	2.60
Dyer's greenweed	1.00	Winter wheat	2.80
Carline thistle	1.00	Winter rapeseed	2.90
Campanula	1.20	Forest trees	5.00–10.00
Broad bean	1.70	Grape vine	12.00–16.00
Spring wheat	1.90	Desert plants	10.00–20.00
Rye	2.00		

Table 2.1.7: Leaf Surface Areas of Selected Plants

Figures from SPECTOR 1956

Species	Leaf surface area (cm^2)	Leaves per plant (n)	Leaf surface area per plant (cm^2)
Apple	18	20,000	318,000
Banana (*Musa acuminata*)	up to 10,000		
Banana (*Musa paradisica*)	2,700–5,200		
Bean	49		
Beech	22	35,000–200,000	4,460,000
Corn (maize)	600–1,320		7,900
Cucumber	29–33		180–1,100
Grape vine	125–150		
Lemon	20–33		
Liriodendron (tulip tree)	130		
Oleander	11		
Potato			17,800
Sunflower	38	59	2,260
Victoria regia	20,000		
Wheat	13–15	5	46–65

Table 2.1.8: Diameter of the Blossoms of Selected Plants

Presented in ascending order.

Data after KERNER V. MARILAUN 1898, MEYER 1964, PERREY 1972, supplemented with several field guides.

Species	Blossom diameter (mm)
Whitlow grass	2.5
Knotweed	3
Lily of the valley	9
Lilac	12
Cyclamen	15
Frostweed	8– 18
European hepatica	15– 30
Tall buttercup	25
Wood anemone	30
Ragged robin	30
Coconut palm (♀)	35
Dog rose	40
Marsh marigold	45
Campanula	50

Species	Blossom diameter (mm)
Hedge false bindweed	30– 60
Iris	50– 70
Poppy (genus)	70
Hibiscus	70
Passion flower (*Passiflora caerulea*)	70– 100
Pumpkin	100– 120
Water lily	120
Opium poppy	160– 180
Cactus (*Cereus grandiflorus*)	200– 220
Magnolia (*Magnolia grandiflora*)	200– 250
Lotus	250
Angel's-tears (*Datura suaveolens*)	250
Victoria regia	300– 400
Rafflesia arnoldii (largest blossom)	1,000 (!)

Table 2.1.9: Specific Weight of Woods

Figures refer to air-dry wood.
Wood consists of 40–50% cellulose, 20–30% hemicellulose, pectin, and sugar, and 20–30% lignin.
Plants are presented in ascending order of specific weight.

After FRANKE 1976, MEYER 1964

Species	Specific weight (g/cm³)	Species	Specific weight (g/cm³)
Balsa wood (the lightest wood)	0.18	Mahogany	0.55
Eastern white pine	0.39	Horse chestnut	0.57
Silver fir	0.41	Platanus	0.63
Redwood (sequoia)	0.42	European rowan	0.64
Douglas fir	0.44	Teak	0.64
Swiss pine	0.44	Birch	0.65
White poplar	0.45	Cherry	0.66
Spruce	0.45	Sweet chestnut	0.66
Pine	0.49	European black pine	0.67
Large-leaf linden	0.49	Sycamore maple	0.67
Common aspen	0.51	Walnut	0.68
Mountain pine	0.51	Elm	0.69
Little-leaf linden	0.52	Robinia	0.70
Black alder	0.53	Pear	0.72
Larch	0.55	Makoré	0.72

(Cont.) →

Species	Specific weight (g/cm³)	Species	Specific weight (g/cm³)
European beech	0.74	Yew	0.84
Norway maple	0.74	Oak	0.86
Plum	0.75	Boxwood (common box)	1.00
Ash	0.75	Ironwood	1.04
European hornbeam	0.77	Ebony	1.10
Hickory	0.80	Guaiacum (or lignum vitae)	
Palisander	0.82	(the heaviest wood)	1.23

Table 2.1.10: Water Content of Selected Plant Organs

Figures are percentage of fresh weight. See also Table 5.2.

Figures from STEWARD 1959, STRASBURGER 1983

Plant species and organ	Water content (%)
Roots	
Barley, top part of plant	93.0
Carrot	88.2
Pine (*P. taeda*), mycorrhiza root	74.2
Pine (*P. taeda*), top part of plant	90.2
Sunflower	71.0
Sprout	
Asparagus, edible tips	88.3
Pine (*P. echinata*). wood	50.0–60.0
Sunflower, entire sprout	87.5
Leaves	
Cabbage	86.0
Corn (maize)	77.0
Lettuce, inner leaves	94.8
Sunflower	81.0
Fruits	
Apple	84.0
Strawberry	89.1
Tomato	94.1
Watermelon	92.1
Seeds	
Barley	10.2
Kernel corn (maize), dry	11.0
Peanut	5.2
Sweet corn	84.8

2.2 Cellular Values

Table 2.2.1: Dimensions of Selected Plant Fibers

After SPECTOR 1956

Type of fiber	Fiber length (cm)	Fiber diameter (µm)	Tensile strength in kg/mm²
Coconut palm	21	18	
Egyptian cotton	3.6	16.4	44
Flax	15– 92	15	76
Hemp	92–183	18–23	84
Jute	150–360	15.5	41
Manila hemp (abaca)	150–350	14–35	40.7
Ramie	20– 30	25.5	91–99
Sisal	45– 90	10–30	40
As a point of comparison:			
Silk	up to 200,000 (= 2 km!)	5–21	45.5
Wool	3.75	25	17.2

Table 2.2.2: Vessel Diameters in Woods

There are sometimes considerable differences between the springwood and summerwood of individual species. On hydraulic conductance in plants, see also Table 2.6.11 and STRASBURGER 1983

Species	Vessel diameter (µm)
Vessel elements	
Alder	20– 90
Ash	120–350
Birch	30–130
Elm	130–140
European beech	16– 80
European hornbeam	16– 80
Horse chestnut	30– 60
Lianas	up to 700 (!)
(e.g. clematis)	
Maple	30–110
Oak	200–300
Pear	40– 80
Robinia	160–400

(Cont.) →

Species	Vessel diameter (µm)
Sweet chestnut	300–350
Tilia (linden, lime)	25– 90
Walnut	120–160
Willow	80–120
Tracheids	
Conifers	20– 40

Table 2.2.3: Number of Stomata on the Leaves of Selected Plants

Figures represent the number of stomata per square millimeter leaf surface area. The number of stomata can vary greatly according to environmental conditions. For instance, leaves in the light present different numbers from their counterparts in the shadows. On the anatomy and function of stomata, see STRASBURGER 1983.

After GREULICH and ADAMS 1962, HELLER 1969, OPPENHEIMER and PINCUSSEN 1929

Species	Upper epidermis	Lower epidermis
Alfalfa	170	140
Apple	0	290
Bean	40	280
Berberis	0	229
Bittersweet (*Solanum dulcamara*)	60	263
Cabbage	140	230
Canadian waterweed	0	0
Cherry	0	250
Corn (maize)	52	68
Cypress spurge	0	259
Eastern white pine	140	0
English holly	0	170
English oak	0	450
European hepatica	15	74
Frostweed	61	194
Geranium	170	140
Iris	65	58
Lilac	0	330
Lily	0	330
Marsh marigold	78	78
Mercurialis	0	65
Military orchid	0	29
Mistletoe	71	75
Net-leaved willow (*Salix retusa*)}	78	134

Species	Upper epidermis	Lower epidermis
Oats	25	23
Olive	0	545
Pea	100	200
Peach	0	225
Poplar	20	115
Potato	50	160
Sea holly (*Eryngium maritimum*)	118	108
Sedum	35	70
Snowdrop	30	45
Spiderworts (*Tradescantie*)	7	28
Sunflower	175	325
Tomato	12	130
Water lily	490	0
Wheat	33	14
Wood anemone	0	67
Woodsorrel	0	35
Yellow Star-of-Bethlehem	27	27

Table 2.2.4: Length of the Stoma Pore in Selected Plants

Figures from OPPENHEIMER and PINCUSSEN 1928, STEWARD 1965

Species	Stoma pore length (µm)
Bishop's goutweed	15
Buckwheat	12
Cactus (*Cereus*)	26
Campanula	19
Corn (maize)	19
Cotton grass	15
Cross-leaved heath	3
Fireweed	9
Hawkweed	12
Marsh marigold	28
Mountain arnica	21
Oak (*Qu. falcata*)	5
Oats	38
Pelargonium	19
Primrose	30
Purple moor grass	10
Runner bean (scarlet runner)	7
Silver ragwort	21

(Cont.) →

Species	Stoma pore length (µm)
Strawberry	10
Sunflower	22
Violet	19
Wheat	38
Wood anemone	38
Woodsorrel	13

Table 2.2.5: Size of Chloroplasts of Selected Plants

Figures represent the largest diameter.

Figures from OPPENHEIMER and PINCUSSEN 1929

Species	Diameter (µm)
Alder	5
Alga (*Vaucheria*)	5
Asarabacca (European wild ginger)	4– 5
Beet (*Beta vulgaris*)	7–10
Canadian waterweed	4– 6
Cuckoopint	5– 7
Daisy	5
Dandelion	5– 6
Ginkgo	3– 5
Lily of the valley	5
Male fern	5
Maple	6
Moss (*Mnium*)	5– 6
Nasturtium	5
Peat moss	3– 4
Pine	5
Pumpkin	5
Shepherd's purse	3– 4
Spiderworts (*Tradescantie*)	3– 5
Stellaria	7–10
Strawberry	4– 5
Tobacco	3– 5
Tulip	5
Walnut	5
Water lily	4– 5
Yew	3– 5

Table 2.2.6: Size of Starch Grains in Selected Useful Plants
After KAUSSMANN 1963, ROHRLICH and BRÜCKNER 1967

Species	Size (μm)
Barley	20– 40
Common bean	max. 60
Corn (maize)	10– 35
Lentil	max. 40
Potato	70–100
Rice	2– 10
Rye	30– 60
Wheat	30– 45

Table 2.2.7: Size of Pollen Grains in Selected Plants

After KERNER V. MARILAUN 1898, MEYER 1964, NILSSON 1977, OPPENHEIMER and PINCUSSEN 1929

Species	Size (μm)
Alder	18–21 × 23– 30
Ash	20–22 × 25– 29
Barley	40– 48
Beech	40–50 × 45– 54
Birch	19–23 × 21– 27
Black poplar	25
Common aspen	28– 34
Common reed	24– 30
Common yarrow	35
Corn (maize)	76–100
Cyclamen	22
Elm	27–33 × 29– 36
European hornbeam	30–34 × 36– 40
Fir	90–110
Forget-me-not	2.5– 3.4
Foxtail grass	36
Hazel	20–25 × 25– 28
Hedge false bindweed	76– 84
Hops	28
Horse chestnut	22–24 × 19– 22
Ivy	23–28 × 26– 31
Lilac	24– 34
Male fern	28–34 × 50– 58
Maple	26–31 × 28– 36
Melon	200–240

(Cont.) →

Species	Size (μm)
Nettle	13–17 × 15– 20
Oak	23– 30
Oats	48– 54
Orchard grass	30– 38
Pine	65– 80
Pumpkin	190–240
Rye	47– 65
Sunflower	39
Tilia (linden, lime)	25–28 × 35– 38
Tobacco	31
Violet	62– 71
Viper's bugloss	10– 14
Water lily	38
Wheat	45– 60
Wild chervil	25–28 × 13– 16
Willow	19–21 × 20– 23
Zigzag clover	43

Table 2.2.8: Dimensions of Spores of a Selection of Well-Known Species of Fungi

Data after MAUCH and LAUBER 1975

Species	Length (μm)	Width (μm)
Basidiomycota		
Bare-toothed russula (also: flirt) (*Russula vesca*)	6.0– 8.0	5.0– 6.0
Bay bolete	11.0–17.0	4.5– 6.0
Chanterelle	7.0– 9.0	4.0– 5.0
Clouded funnel or agaric (*Clitocybe nearis*)	6.0– 7.0	3.0– 4.0
Common earthball	8.0–12.0	Spherical
Death cap	8.0–11.0	Spherical
Dry rot fungus (*Serpula lacrymans*)	8.5–11.0	4.5– 6.0
Field mushroom	7.0– 8.0	4.0– 5.0
Fly agaric	9.0–11.0	6.0– 9.0
Honey fungus	7.0– 9.0	5.0– 6.0
Larch bolete	8.0–11.0	3.0– 4.0
Livid pinkgill (*Entoloma sinuatum*)	8.0–10.0	7.0– 8.5
Orange oak bolete (*Leccinum quercinum*)	13.0–17.0	4.0– 5.0
Oyster mushroom	8.0–12.0	3.0– 4.0
Panther (cap) (*Amanita pantherina*)	10.0–12.0	7.0– 8.0
Parasol	15.0–20.0	10.0–13.0
Peppery milk cap (*Lactarius piperatus*)	8.0– 9.5	5.5– 7.0
Porcini	13.0– 17.0	3.5– 6.0

Species	Length (μm)	Width (μm)
Ramaria aurea (coral fungus)	8.0–15.0	3.0– 6. 0
Red-staining inocybe	9.0–14.0	5.0– 8.0
Sarcodon imbricatus	6.0– 7.0	5.0– 6.0
Shaggy ink cap (*Coprinus comatus*)	12.0–16.0	7.0– 8.0
Sheathed woodtuft		
(*Kuehneromyces mutabilis*)	6.0– 7.0	3.0– 4.5
Stinkhorn (*Phallus impudicus*)	3.5– 5.0	1.5– 2.5
Sulphur tuft	6.0– 8.0	4.0
Tinder polypore	15.0– 20.0	5.0– 7.0
Ascomycota		
Black morel	18.0– 24.0	10.0–14.0
Ergot	100.0	1.0
False morel	17.0– 25.0	8.0–12.0

Table 2.2.9: Chromosome Sets of Selected Plants

Figures given refer to the diploid set. For diploid values for animals, see Table 1.9.3.

After BRESCH and HAUSMANN 1972, GÖTZ and KONRAD 1977, HESS, LANDOLT and HIRZEL 1967, NAGL 1976, NIGON and LUEKEN 1976, STARCK 1975

Species	Chromosomes (*n*)	Species	Chromosomes (*n*)
Adder's-tongue	480	Campanula	34
Alder	28	Canadian waterweed	24
Alga (*Cladophora*)	32	Caraway	20
Almond	16	Carrot	18
Apple	34, 51, 68	Chamomile	18
Ash	46	Cherry, sour	32
Asparagus	20	Cherry, sweet	16, 24, 32
Autumn crocus	38		64, 144
Banana	22, 33, 44	Christmas rose	32
Barley	14	Clusterhead	30
Bedstraw, hedge	22	Coconut	32
Bee orchid	36	Coltsfoot	60
Birch	84	Common bean	22
Brackenfern	104	Common yarrow	36, 54
Burdock	36	Corn (maize)	20
Bush vetch	14	Corn poppy	14
Button mushroom	8	Cornflower	24

(Cont.) →

Species	Chromosomes (n)	Species	Chromosomes (n)
Cross-leaved heath	24	Pear	68, 85
Cucumber	14	Peppermint	68, 72
Cyclamen	48, 84	Pine	24
Cypress spurge	20	Plum	48
Dandelion	24	Poplar	38
Eelgrass	12	Potato	48
English oak	24	Primrose	22
Euglena	approx. 200	Privet	46
European beech	24	Pumpkin	40
European rowan	34	Rape	38
European spindle	64	Red clover	14
Field or hedge maple	26, 78	Sedge, hairy	112
Fir	24	Shepherd's purse	16, 32
Forget-me-not	22	Snowdrop	24
Ginkgo	24	Sorrel ♀	14
Gorse	96	♂	15
Heath	16	Spikemoss (*Selaginella*)	18
Hellebore	32	Spinach	12
Horse chestnut	40	Spring gentian	28
Horsetail	216	Spruce	24
Juniper	22	Strawberry, garden	14, 28, 42, 56
Lady's slipper	20		
Larch	24	Strawberry, woodland	14
Lilac	46–48	Sundew	20
Little-leaf linden	72	Sunflower	34
Male fern	164	Tobacco	48
Marsh marigold	32, 48, 56	Tomato	24
Meadow sage	18	Vaccinium	24
Mistletoe	20	Valerian	18
Mustard	24	Venus flytrap	30
Northern firmoss	68	Violet	20
Oats	42	Wheat	
Onion	16	Einkorn	14
Opium poppy	22	Durum wheat, emmer	28
Oxeye daisy	36	Soft wheat, spelt	42
Pansy	26	Wild radish	18
Parsley	23	Willow	38
Pasque flower	32	Wine grape	38, 57
Pea	14		76
Peach	16	Yew	24
Pear	34, 51		

2.3 Germination and Growth

Table 2.3.1: Germination Time and Optimal Germination Temperature of the Seeds of Selected Plants

A range given for the germination temperature indicates that the plants prefer daily variation (16 h of the upper and 8 h of the lower temperature). On germination speed, see Table 2.3.2.

Figures after ALTMAN and DITTMER 1972

	Germination time (days)	Germination temperature (°C)
Asparagus	7–21	20–30
Barley	4–7	15
Broad bean	4–14	20
Carrot	6–21	20–30
Cauliflower	3–10	20–30
Corn (maize)	4–7	25
Cornflower	4–8	15
Dandelion	7–21	20–30
Forget-me-not	5–12	20
Hemp	3–7	20–30
Larkspur	10–21	15
Lentil	7–12	20
Lettuce	7	20
Lupin	4–10	20
Meadow sage	4–12	20–30
Nasturtium	14	18
Oats	5–10	15
Onion	6–10	20
Parsley	11–28	20–30
Pea	5–8	20
Pumpkin	4–7	20–30
Radish	4–6	20
Red beet	3–10	20–30
Rye	4–7	15
Spinach	7–21	10
Sunflower	3–7	20–30
Sweet william	8	20
Tobacco	7–14	20–30
Tomato	5–14	20–30
Violet	12	20–30
Wheat	4–7	15

Table 2.3.2: Germination Speed of the Seed of Selected Useful Plants

Figures given represent the number of hours until the radicles become visible. Since this is highly temperature dependent, values are provided for 10, 15, and 25 °C.

After MEYER 1964

	10 °C	15 °C	25 °C
Alfalfa	90	66	48
Barley	72	48	72
Carrot	162	102	56
Corn (maize)	272	78	56
Cucumber	–	–	96
Oats	90	66	48
Pea	72	42	–
Poppy (genus)	114	60	48
Pumpkin	–	258	52
Rape	48	24	24
Red clover	72	42	24
Rye	54	30	32
Sunflower	600	72	32
Wheat	84	48	33

Table 2.3.3: Vernalization Temperatures and Periods of Selected Plants

Vernalization is defined as the temporary effect of low temperatures on seeds or plants in initiating the flowering process. In many useful plants, there are great differences between individual varieties.

Figures after various authors in ALTMAN and DITTMER 1973

Species and plant part	Period (days)	Optimal temperature (°C)
Bean, seeds	30–35	2
Beet, plants	30–50	0–2
Beet, seeds	80	0–5
Cabbage, seeds	28	0
Carrot, plants	40–65	4
Cauliflower, plants	14	5
Celery, plants	15–30	4,5–10
Common evening primrose, seeds	30	4
Corn (maize), seeds	34	3
Crimson clover, seeds	40	0
Dactylis glomerata, plants	30–90	0–2
Endive, seeds	28	1–2.5

Species and plant part	Period (days)	Optimal temperature (°C)
Festuca elatior, plants	30–90	0–2
Kohlrabi, plants	60–90	5
Lentil, seeds	10–12	6 –10
Lettuce, seeds	10–20	2 – 5
Lolium perenne, seeds	0–90	0 – 5
Lupin, seeds	14–21	2 – 5
Oats, seeds	30–45	0–2
Opium poppy, seeds	35	2 – 3
Pea, seeds	20–30	2–7
Radish, seeds	10–46	0 – 5
Spinach, seeds	10–15	2 – 8
Spring barley, seeds	0–15	6 – 8
Spring rye, seeds	0–14	0 – 5
Spring wheat, seeds	0–14	0 – 8
Sunflower, seeds	35	2
Sweet william, plants	42–63	5
Trifolium pratense, seeds	10–40	3 – 8
White mustard, seeds	10–30	0 – 3
Winter barley, seeds	20–40	0 – 3
Winter rye, seeds	30–50	0 – 5
Winter wheat, seeds	40–70	0 – 3

Table 2.3.4: Growth Speed of Selected Plant Organs Under Optimal Conditions

The growth of plants is heavily temperature dependent; on this see Table 2.3.11. Most plants grow at a speed of less than 0.005 mm/min.

Figures after MEYER 1964, OPPENHEIMER and PINCUSSEN 1929

Plant species and organ	Growth speed			
Asparagus, shoot	0.08	mm/min =	11.1	cm/day
Autumn crocus, pollen tubes	0.015	mm/min =	2.16	cm/day
Bamboo, shoot	0.4	mm/min =	57	cm/day
Banana, leaf sheath	1.1	mm/min =	160	cm/day
Broad bean, radicle	0.012	mm/min =	1.7	cm/day
Bryony, shoot	0.056	mm/min =	8	cm/day
Hemp, sprout	0.05	mm/min =	7.2	cm/day
Ink cap, fruiting body stalk	0.22	mm/min =	31.4	cm/day
Mold (*Mucor*), hyphae	0.026	mm/min =	3.74	cm/day
Oats, coleoptile	0.025	mm/min =	3.7	cm/day
Pea, root	0.02	mm/min =	2.88	cm/day

(Cont.) →

Plant species and organ	Growth speed			
Pumpkin, shoot	0.1	mm/min =	14.4	cm/day
Rye, filaments				
(max. unwinding time, 10 min)	2.5	mm/min		
"Veiled" mushroom				
(*Dictyophora*), fruiting body				
(max. unwinding time, 15 min)	5	mm/min		

Table 2.3.5: Annual Vertical Growth Over Lifespan of Selected Important Harvested Trees Under Optimal Conditions

For maximum sizes, see Table 2.3.6; for maximum ages, Table 2.1.3; and for rotation periods, Table 2.1.4.

After Bertelsmann 1979, Meyer 1964

	1–20	21–40	41–60	61–80	81–100	101–120 years	
Fir	12.5	54.0	48.0	27.5	20.5	11.0	cm
Spruce	30.5	48.0	38.0	25.0	16.0	11.5	cm
Pine	44.5	40.0	27.0	19.0	19.0	11.5	cm
Beech	27.5	40.0	34.0	27.0	14.0	13.5	cm
Oak	46.5	45.0	19.5	19.5	17.0	10.5	cm

Table 2.3.6: Height of Selected Trees

For trunk circumferences, see Table 2.3.8; for trunk diameters, Table 2.3.7; and for slenderness ratio, see Table 2.3.9.

Data after Bertelsmann 1979, Hofmeister 1977, Meyer 1964

Species	Normal height (m)	Maximum height (m)
Ash	20– 40	50
Australian mountain ash		
(*Eucalyptus regnans*)	100–110	152
Black alder	10– 25	30
Black poplar	15– 30	35
Cedar	25– 30	40
Coconut palm	25– 30	32
Cypress (*Cupressus*)	20– 50	52
Douglas fir	10– 25	25
English oak	20– 40	50

Species	Normal height (m)	Maximum height (m)
European beech	15– 40	44
European hornbeam	5– 25	28
Fern trees	10	22
Field or hedge maple	10– 15	22
Larch	30– 40	53
Large-leaf linden	15– 30	35
Pine	20– 40	48
Redwood (sequoia)	100	132
Robinia	10– 25	25
Silver fir	30– 50	75
Spruce	30– 50	60
Sycamore maple	15– 30	40

Table 2.3.7: Maximum Trunk Diameters of Selected Tree Species

Presented in ascending order of size.
For maximum tree heights, see Table 2.3.6; for maximum tree circumferences, see Table 2.3.8; and for tree slenderness ratios, see Table 2.3.9.

After BERTELSMANN 1979, MEYER 1964

Species	Maximum trunk diameter (m)
Field or hedge maple	0.7
Birch	0.8
Juniper	0.9
European hornbeam	1.0
Pine	1.0
Larch	1.6
Ash	1.7
Spruce	2.0
European beech	2.0
Fir	3.0
Norway maple	3.0
Elm	3.0
Yew	4.9
Large-leaf linden	9.0
Redwood (sequoia)	11.0
Baobab	15.0

Table 2.3.8: Greatest Recorded Trunk Circumferences for Selected Trees

Figures represent individual trees with known locations that are provided here. Presented in ascending order.
For maximum tree heights, see Table 2.3.6; for maximum trunk diameters, see Table 2.3.7; and for slenderness ratios, see Table 2.3.9.

After MEYER 1964

Species	Circumference (m)	Location
Ivy	1.80	Gignac/France
English elm	12.45	Schimsheim/Germany
Oak	13.00	Newland/England
Tilia (linden. lime)	17.00	Staffelstein/Germany
Yew	18.00	Braburn/England
Sweet chestnut	20.00	Italy
Bald cypress	31.00	Santa Maria el Tule/Mexico
Redwood (sequoia)	35.00	Mariposa/USA
Platanus	47.00	Böjuk-Dere/Turkey
Baobab	47.00	Tanganyika/Africa

Table 2.3.9: Slenderness Ratio and Height of Selected Plants

The slenderness ratio is the quotient of the height of a plant divided by its diameter at the base. As a point of comparison, an architectural structure: the 211-meter Stuttgart television tower has a slenderness ratio of 19.
The slenderness ratio is only comparable in plants of the same absolute height, since in larger plants with the same slenderness ratio, their bending strength decreases.
Tree heights are also provided in Table 2.3.6.
Species are listed in ascending order, according to their slenderness ratio.

Values after HAGER 1960, HEYNERT 1976, NACHTIGALL 1977 and 1979

Species	Height (m)	Maximum slenderness ratio
Baobab	20.0	2.5
Giant sequoia	135.0	11
Evergreen redwood (sequoia)	100.0	15
Cedar	40.0	20
Eucalyptus	128.0	28
Fir	70.0	42
Spruce	60.0	60
Palm	30.0	60
Sunflower	4.0	100
Bamboo	40.0	133
Sugarcane	6.0	200
Rye	1.5	500

Table 2.3.10: Length of Runners of Selected Plants Over the Course of a Year

In ascending order, according to length.

Data after KERNER V. MARILAUN 1898

Species	Length (cm)
Yellow mountain saxifrage	4
Yellow anemone	5– 10
Oregano	5– 10
Lemon balm	5– 10
Common yarrow	10– 15
Field horsetail	10– 15
Sweet violet	13
Dentaria glandulosa	15– 20
Sand sedge	15– 20
Spearmint	15– 20
Common bugle	20
Woodruff	20– 25
Peppermint	20– 25
Common madder	20– 25
Mercurialis	25– 30
Quackgrass	25– 30
Bishop's goutweed	25– 30
Stiff clubmoss	30– 40
Field bindweed	30– 55
Soapwort	30– 55
Nettle	35– 45
Hairy sedge	35– 45
Sea milkwort	45– 55
Creeping avens	50
Broad-leaved ragwort	55– 60
Creeping buttercup	60
Coltsfoot	60– 75
Canada goldenrod	60– 75
Rosebay willow herb	85–100
Pestilence wort	100–150
Silverweed	110
Ground ivy	126
Creeping cinquefoil	130
Stone bramble	140
Strawberry	150
Periwinkle	200

Table 2.3.11: Cardinal Temperature Points for the Growth of Selected Useful Plants

Plant growth only occurs at a temperature between the minimum and maximum.

After BROUWER 1972, ELLENBERG 1978, GESSNER 1942, MADLE 1967, STILES and COCKING 1969, WALTER 1962

Species	Minimum (°C)	Optimum (°C)	Maximum (°C)
Alfalfa	1	30	37
Barley	3–4	20	30
Bean	10	32	37
Beech	− 5	20–22	43
Corn (maize)	8–10	34	46
Cucumber	15	35	50
Fir	− 7	14–20	38
Hemp	0– 5	37–44	44–50
Lentil	2		28
Maple (*A. platanoides*)	7		28
Melon	15–18	31–37	44–50
Oats	4– 5	25	30
Pine	7		34
Pumpkin	10–15	37–44	44–50
Rye	2– 3	25	30
Spruce	− 7	12–18	36
Sunflower	5–10	31–37	37–44
Tobacco	13	28	35
Wheat	5– 6	29	40

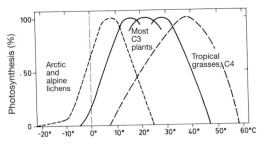

Fig. 2.3.11: Dependence of the position of the cardinal points of net photosynthesis upon temperature in selected types of plants. See also Table 2.3.11.
From STRASBURGER 1983, p. 270

Table 2.3.12: Optimal pH Soil Levels for Plant Growth

Some specialists can populate soils with extreme pH levels. For instance, peat mosses (sphagnum) and cotton grass still grow at pH levels of 3.3-3.5. The lowest recorded pH value, of 1.6, was set by a Central American species of spikerush (*Eleocharis schaffneri*).

Data after various authors from ALTMAN and DITTMER 1973, LÖTSCHERT 1969

Species	pH optimum
Asparagus	6.0–8.0
Bilberry	4.5–6.0
Carrot	5.5–7.0
Common aspen	4.5–5.5
Corn (maize)	5.5–7.5
English holly	5.5–6.5
European beech	6.0–7.5
Fir	4.5–6.5
Ginkgo	5.5–7.0
Horse chestnut	5.5–7.0
Larch	4.5–7.5
Melon	6.0–8.0
Oak	6.0–7.5
Oats	5.0–6.8
Onion	6.0–7.5
Parsley	5.0–7.0
Pea	6.0–8.0
Pine	4.5–6.5
Potato	5.4–6.7
Pumpkin	5.5–7.0
Rhododendron	4.5–6.0
Rose	5.5–7.0
Rye	5.0–7.0
Sorrel	4.0–7.5
Spruce	4.5–6.5
Strawberry	5.0–6.5
Sunflower	6.0–7.5
Tilia (linden, lime)	6.0–7.5
Tobacco	5.5–7.5
Violet	6.0–7.5
Walnut	6.0–7.5
Wheat	5.5–7.5
Wine grape	6.0–8.0

Table 2.3.13: Polar Limits for the Cultivation of Selected Useful Plants

Figures are in degrees of northern latitude. As a point of reference, the latitudes of several important cities are also provided. Since crop cultivation is also heavily dependent on the amount of precipitation, see also Table 2.3.14.

After ANDREAE 1976

Plant	Northern boundary	Comparative approx. location of selected cities (and other geographical entities)
Coconut palm	15°	15° Manila, Dakar, Guatemala
Oil palm	16°	
Sisal	19°	19° Mexico City, Mumbai (Bombay)
		21° Mecca, Hanoi
Cacao, coffee	22°	22° Kolkata (Calcutta)
Banana, cassava	23°	23° Havana
Caoutchouc	25°	
		26° Miami
		30° Cairo
		32° Tel Aviv
Sweet potato	35°	
Cotton	38°	38° Athens, Palermo, San Francisco
Sugarcane	39°	
		40° Madrid
Peanut, tea	41°	41° Naples, New York
Citrus	42°	42° Rome
Soy bean, millet, olive	45°	45° Turin, Bordeaux
		48° Vienna, Munich
		49° Paris, United States' northern border
		50° Mainz, Winnipeg
Wine grape	51°	51° London
Rice	52°	52.5° Berlin
Tobacco	53°	
Kernel corn	54°	
		56° Moscow
		59° Stockholm
		60° St. Petersburg, Oslo
Beetroot	61°	
Wheat	63°	63° Trondheim
		69° Murmansk
Barley, potatoes	70°	

Table 2.3.14: Dryness Limits for the Cultivation of Selected Useful Plants

Figures represent the minimum required annual precipitation in millimeters. As a point of reference, figures for the annual precipitation of several well-known locations in different climate zones are also provided. Since crop cultivation is also highly dependent on temperature, see also Table 2.3.13.

Data after ANDREAE 1976

Species	Required annual precipitation (mm)	Annual precipitation of selected cities	
		Goa	2,506
		Mumbai (Bombay)	1,813
Caoutchouc, oil palm	1,500	New Orleans	1,545
Tea, yams	1,500		
Sugarcane	1,400		
Cacao, coconut palm	1,300	Chennai (Madras)	1,241
		Shanghai	1,120
		Milan	1,011
Coffea arabica	900	Munich	935
Rice	800	Rome	836
Kernel corn	760	Lisbon	754
		Algiers	728
		Palermo	644
		Vienna	644
		Frankfurt	604
		Adana	600
		Berlin	581
		Moscow	535
Sweet potato, cotton	500	Magdeburg	508
Cassava, tobacco	500		
Sugar beet	450	Los Angeles	440
Potatoes, sesame	400		
Peanut, wheat	300		
Sisal, barley	250		
Olive	200	Phoenix (Ariz.)	206
		Bagdad	145

2.4 Reproduction: Blossoms, Fruits, Seeds

Table 2.4.1: Linnaeus' Flower-Clock

The times apply to Uppsala, Sweden (60° north latitude).

After MEYER 1964

Time	The following blossoms open:	The following blossoms close:
3–5 a.m.	Yellow goat's beard	
4–5 a.m.	Chicory, bristly oxtongue	
5 a.m.	Tawny daylily, common sow thistle, Icelandic poppy	
5–6 a.m.	Dandelion	
6 a.m.	Narrowleaf hawkweed Spotted cat's-ear	
6–7 a.m.	Bladderpod, wall hawkweed, Common hawkweed	
7 a.m.	*Anthericum ramosum*, white water lily	
8 a.m.	Scarlet pimpernel, tunic flower	
8–10 a.m.		Dandelion
9–10 a.m.	Field marigold, Red sandspurry	Yellow goat's beard
10 a.m.		Chicory
11–12 a.m.		Common sow thistle
12 noon		Field marigold
1 p.m.		Tunic flower
2 p.m.		Common wall hawkweed Hawkweed
2–3 p.m.		Red sandspurry
3–4 p.m.		*Anthericum ramosum*
4 p.m.		Bladderpod
4–5 p.m.		Spotted cat's-ear
5 p.m.		Narrowleaf hawkweed, white water lily
7 p.m.		Icelandic poppy
7–8 p.m.		Tawny daylily

Table 2.4.2: Floral Formulas of Selected Plant Families

The table presents the floral formula typical for a given plant family. Where distinct formulas exist within one family, these are also presented here for important species; the same applies for exceptions in well-known species. The table focuses mainly on important European plant families.

A = Androecium (whorl of stamens), Ca = calyx (sepal whorl), Co = corolla (petal whorl), G = gynoecium (carpel or carpels), P = perigone (where corolla and calyx are indistinguishable); n = numerous. Parentheses indicate that the parts referred to are fused.

Data after GRAF 1975, SCHMEIL and FITSCHEN 1982, STRASBURGER 1983

Family		P	Ca	Co	A	G
Aster (*Asteraceae* and *Cichoriaceae*)				(5)	(5)	(2)
Beech (*Fagaceae*)						
E.g., European beech	♂	5–6				(4)–(15)
	♀	3+3				(3)
Birch (*Betulaceae*)						
Birch	♂	2+2			2	
	♀					(2)
Hazel	♂				4	
	♀					(2)
European hornbeam	♂				4-12	
	♀					(2)
Bluebells (*Campanulaceae* family)			5	(5)	5	(3)
Borage (*Boraginaceae*)			5	(5)	5	(2)
Buttercup (*Ranunculaceae*)						
Anemone sp.		3+3			n	n
Ranunculus sp.			5	5	n	n
Carrot (*Apiaceae*)			5	5	5	(2)
Duckweed (*Lemnaceae*)	♂				1	
	♀					1
Figwort (*Scrophulariaceae*)			(5)	(5)	5	(2)
Veronica sp.			(5)	(5)	2	(2)
Gentian (*Gentianaceae*)			5	5	5	(2)
Geranium (*Geraniaceae*)			5	5	(5)–(5+5)	(5)
Goosefoot (*Chenopodiaceae*)		5			5	(2)
Honeysuckle (*Caprifoliaceae*)			5	5	5	(2)–(3)
Heath (*Ericaceae*)						
Erica, Calluna spp.			4	(4)	8	(4)
Vaccinium sp.			5	(5)	10	(5)
Ivy (*Araliaceae*)			5	5	5	(3)–(5)

(Cont.) →

Family		P	Ca	Co	A	G
Knotweed (*Polygonaceae*)						
Rheum sp.		6			9	(3)
Rumex sp.		6			6	(3)
Lily (*Liliaceae*)		3+3			3+3	(3)
Paris, Maianthemum spp.		4	4	4+4	(4)	
Linden (*Tiliaceae*)			5	5	n	(5)
Mallow (*Malvaceae*)						
e.g., *Malva* sp.			5	5	n	(n)
Maple (*Aceraceae*)			5	5	8	(2)
Mint (*Lamiaceae*)			(5)	(5)	4	(2)
Mustard (*Brassicaceae*)			4	4	4+2	(2)
Nettle	♂	4			4	
(*Urticaceae*)	♀	2–5				1
Nightshade (*Solanaceae*)			(5)	(5)	5	(2)
Orchid (*Orchidaceae*)		6			1–2	(3)
Orpine (*Crassulaceae*)			5	5	10	5
Pea or legume (*Fabaceae*)						
Laburnum sp.			(5)	5	(10)	1
Vicia sp.			(5)	5	(9)+1	1
Pink (*Caryophyllaceae*)			(5)	5	5+5	(5)
Plantain (*Plantaginaceae*)			4	(4)	4	(2)
Primrose (*Primulaceae*)			(5)	(5)	5	(5)
Poppy (*Papaveraceae*)			2	2+2	n	(2)
Rose (*Rosaceae*)			5	5	5-?	1-?
Rush (*Juncaceae*)		3+3			3+3	(3)
Saxifrage (*Saxifragaceae*)			5	5	5+5	(2)
Sedge (*Cyperaceae*)						
Scirpus sp.		3+3			3	(3)
Carex sp.	♂				3	
	♀ ♂					(2)–(3)
Spurge	♂				1	
(*Euphorbiaceae*)	♀					(3)
Sundew (*Droseraceae*)			5	5	5	(3)
Teasel (*Dipsacaceae*)			4–5	(4)–(5)	4	1
True grass (*Poaceae*)		(2)+2			3	(3)
Violet (*Violaceae*)			5	5	5	(3)
Water lily (*Nymphaeaceae*)			4	n	n	n
Willow (*Salicaceae*)						
Salix sp.	♂				2	
	♀ ♂					(2)
Populus sp.	♂				3–30	
	♀					(2)
Wood sorrel (*Oxalidaceae*)			5	5	5+5	5

Table 2.4.3: Blossoming Period of Selected Plants

Figures represent the lifespan of an individual blossom.
Presented in ascending order, according to length.

Data after KERNER V. MARILAUN 1898, MOLISCH 1965

Species	Blossoming period (days)	Species	Blossoming period (days)
Common evening primrose	1	Yellow daylily	6
Angel's-tears	1	Pelargonium	7
Meadow cranesbill	2	Tall buttercup	7
Hill willow-herb	2	Marsh grass of Parnassus	8
Opium poppy	2	European hepatica	8
Veronica aphylla	2	Saxifrage (saxifraga bryoides)	8
Field rose	2	Winter aconite	8
Wild mustard	2	Cyclamen	10
Common agrimony	3	Prickly saxifrage	12
Alpine rockrose	3	Saffron crocus	12
Dwarf honeysuckle	3	Cranberry	18
Pearlwort (sagina)	4	Tropical orchids	
Red campion	4	Cattleya labiata	30
Sedum atratum	4	Cypripedium insigne	40
Snake's head fritillary	5	Phalaenopsis grandiflora	50
Flax (Linum viscosum)	5	Oncidium cruentatum	60
Common centaury	5	Cypripedium villosum	70
Purple foxglove	6	Odontoglossum rossii	80
Branching centaury	6		

Table 2.4.4: Critical Day Length for Inducing Blossoming in Long-Day and Short-Day Plants

Figures from GREULACH and ADAMS 1962, JENSEN and SALISBURY 1972. See also STRASBURGER 1983

Species	Critical day length (h)
Long-day plants	
Bent grass (Agrostris palustris)	< 16
Brome (Bromus inermis)	< 12.5
Dill	< 11
English rye-grass (Lolium perenne)	< 13
Henbane	< 10
Hibiscus	< 12
Italian rye-grass (Lolium italicum)	< 11
Oats	< 9
Orchard grass	< 12

(Cont.) →

Species	Critical day length (h)
Radish	< 12
Red clover	< 12
Reed canary grass (*Phaleria arundinacea*)	< 12.5
Rudbeckia	< 10
Spinach	< 13
Timothy grass (*Phleum nodosum*)	< 14.5
Timothy grass (*Phleum pratense*)	< 12
Wheatgrass (*Agropyron smithii*)	< 10
Winter barley	< 12
Winter wheat	< 12
Short-day plants	
Bryophyllum	> 12
Chrysanthemum	> 15
Cosmos	> 14
Kalanchoe	> 12
Poinsettia (*Euphorbia pulcherrima*)	> 12
Redwood (sequoia)	> 14
Strawberry (*Fragaria chiloensis*)	> 10
Tobacco	> 14
Violet (*Viola papilionacea*)	> 11
Winter rice	> 12

Table 2.4.5: Blossoming Data for Selected Trees and Shrubs Native to Germany

For data on monocotyledonous plants, see Table 2.4.6; for true grasses (*Poaceae*), see Table 2.4.7; and for dicotyledonous plants, see Table 2.4.8. Please note that in northern Germany, blossoms (e.g., of the apple tree) start 3 weeks later than in the Rhine valley or at Lake Constance.

Data from FLINDT 1985, SCHMEIL and FITSCHEN 1968

Species	Months	Species	Months
Alder	III–IV	Bog rosemary	V–VI
Alder buckthorn	V–VI	Boxwood (common box)	III–IV
Apple	IV–VI	Clematis	VI–IX
Apricot	III–IV	Common aspen	II–IV
Ash	V–VI	Common barberry	V–VI
Basket willow	III–IV	Cornelian cherry	III–IV
Birch	IV–V	Cowberry (lingonberry)	V–VII
Bird cherry	IV–VI	Cross-leaved heath	VII–VIII
Black poplar	III–IV	Crowberry	V–VI
Blackberry	VI–VII	Dictamnus	V–VI
Blackthorn	III–IV	Dog rose	VI

Species	Months	Species	Months
Dog rose	VI	Pear	IV–V
Douglas fir	IV–V	Platanus	V
Eared willow	IV–V	Plum	IV–V
Elderberry	VI–VII	Privet	VI–VII
English elm	III–IV	Quince	V–VI
English holly	V–VI	Raspberry	V–VI
European beech	IV–V	Rhododendrons	V–VII
European hornbeam	IV–VI	Robinia	V–VI
European rowan	V–VII	Rugosa rose	VI–VII
European spindle	V–VII	Scotch broom	V–VI
False medlar	IV–VI	Scotch pine	V–VI
Field or hedge maple	V	Siberian dogwood	V–VI
Forsythia	III–IV	Silver fir	V–VI
Goat willow	III–IV	Sour cherry	IV–V
Gooseberry	IV–V	Spiny restharrow	VI–VII
Hawthorn	V–VI	Spruce	IV–VI
Hazel	II–IV	Sweet chestnut	VI
Horse chestnut	IV–V	Sweet gale	IV–V
Ivy	IX–X	Swiss pine	VI–VII
Juniper	IV–V	Sycamore maple	IV–VI
Laburnum	IV–VI	Vaccinium	V–VI
Larch	IV–VI	Walnut	IV–V
Large-leaf linden	VI	Wayfaring tree	IV–V
Lilac	V–VI	White poplar	III–IV
Little-leaf linden	VI–VII	Whitebeam	V–VI
Mezereon	II–IV	Wild cherry	IV–V
Mistletoe	III–V	Wild rosemary	V–VI
Mugo pine	V–VI	Wine grape	VI–VII
Norway maple	IV–V	Winged broom	V–VII
Oak	IV–V	Yew	III–IV
Peach	IV–V		

Table 2.4.6: Blossoming Data for Selected Monocotyledonous Plants Native to Germany

Grasses not included here; for these, see Table 2.4.7. For figures on trees and shrubs, see Table 2.4.5; and for dicotyledonous plants, see Table 2.4.8.

Figures from FLINDT 1985, SCHMEIL and FITSCHEN 1968

Species	Months in bloom	Species	Month in bloom
Alpine squill	III–IV	Lesser rattlesnake plantain	VII–VIII
Anthericum	VI–VIII	Lily of the valley	V–VI
Arrowhead	VI–VIII	Martagon lily	VII–VIII
Asparagus	V–VI	Military orchid	V–V
Autumn crocus	VIII–IX	Orchid	
Bird's-nest orchid	V–VI	Bee	V–VII
Branched bur-reed	VI–VIII	Fly	V–VI
Broadleaf cattail	VI–VII	Late spider	V–VI
Canadian waterweed	V–VIII	Pondweed	
Common spikerush	VI–VII	Curly	V–IX
Common spotted orchid	V–VIII	Floating leaf	V–VIII
Cotton grass/sedge		Opposite-leaved	VI–VII
Common	III–V	Ramsons	V–VI
Sheathed	III–IV	Rannoch rush	V–VII
Crocus	III–IV	Round-fruited rush	VI–IX
Cuckoopint	IV–V	Sea club-rush	VII–VIII
Daffodil	III–IV	Sedge	
Early purple orchid	V–VI	Common	V–VIII
Eelgrass	VI–X	Hairy	IV–VI
False lily of the valley	IV–VI	Sand	V–VI
Flowering rush	VI–VIII	Star	V–VI
Fragrant orchid	V–VIII	Tufted	IV–V
Frog orchid	V–VI	White	V–VIII
Frog's bit	V–VIII	Snake's head fritillary	IV–V
Great woodrush	IV–V	Snowdrop	II–IV
Greater duckweed	V–VI	Solomon's seal	V–VI
Hellebore	VI–VIII	Spring snowflake	II–IV
Helleborine		Star of Bethlehem	
Broad-leaved	VI–IX	(*Gagea villosa*)	III–V
Marsh	VI–VIII	Twayblade	V–VII
Herb Paris	V	Water arum	V–VII
Lady's slipper	V–VII	Water plaintain	VI–IX
Lesser butterfly orchid	V–VII	Yellow flag	V–VI

Table 2.4.7: Blossoming Data for Selected True Grasses (*Poaceae*) Native to Germany

Data after FLINDT 1985, SCHMEIL and FITSCHEN 1968

Species	Months	Species	Months
Alkali grass	VI–IX	Orchard grass	V–VI
Barley	VII	Purple moor grass	VII–IX
Bearded darnel	VI–VIII	Quackgrass	VI–VII
Brome		Quaking grass	V–IX
Upright	V–VII	Redtop	VI–VIII
Soft	V–VII	Reed canary grass	VI–VIII
Common reed	VII–IX	Rye	V–VI
Common velvet grass	VI–VIII	Smooth meadow-grass	
Cordgrass	VII–IX	(Kentucky bluegrass)	V–VII
Crested dog's-tail	VI–VII	Sweet vernal grass	V–VIII
European beach or		Tall oatgrass	VI–VII
marram grass	VI–VII	Timothy	V–VII
Feather grass	VII–VIII	Tufted hairgrass	VII–IX
Fescue (*festuca varia*)	VII–VIII	Wheat	VI
Meadow	VI–VII	Wild barley	VI–XI
Loose silky-bent	VI–VII	Wild oat	VI–VIII
Mat grass	V–VII	Wood manna-grass	VI–VII
Millet grass	V–VII	Wood melick	V–VI
Oats	VI–VII	Wood small reed	V–VIII

Table 2.4.8: Blossoming Data for Selected Dicotyledonous Plants Native to Germany

Trees and shrubs are excluded here; for these, see Table 2.4.5. For data on monocotyledonous plants, see Table 2.4.6, for true grasses, see Table 2.4.7.

Data after FLINDT 1985, SCHMEIL and FITSCHEN 1968

Species	Months in bloom	Species	Months in bloom
Anemone, yellow	III–V	Buckbean	V–VI
Arnica	V–VIII	Bugle, common	V–VIII
Asarabacca		Bull thistle	VIII–X
(European wild ginger)	III–V	Burdock	VII–IX
Belladonna	VI–VIII	Burnet, Great	VI–IX
Bishop's goutweed	V–IX	Salad	V–VI
Biting or gold-moss stonecrop		Buttercup, Corn	V–VII
(*Sedum acre*)	VI–VIII	Creeping	V–VIII
Black medick	V–IX	Tall	V–X
British yellowhead	VI–IX	Campion	
Bryony	VI–IX	Red	IV–IX

(Cont.) →

Species	Months in bloom	Species	Months in bloom
White	VI–IX	Cyclamen	VI–IX
Caraway	V–VII	Cypress spurge	IV–VII
Carline thistle	VII–IX	Daisy	III–XI
Carrot, wild	V–VII	Dandelion	IV–VII
Centaury	VII–IX	Dead-nettle	
Chicory	VII–IX	Red	III–X
Christmas rose	II–VI	White	IV–VIII
Cleavers	V–X	Dodder	VI–IX
Clusterhead		Dog fennel or chamomile (*Anthemis*)	
or carthusian pink	VI–VIII		VI–IX
Coltsfoot	III–IV	Dog violet	IV–VI
Comfrey	V–VII	Edelweiss	VII–IX
Common agrimony	VI–VIII	Enchanter's nightshade	VI–IX
Common alkanet or bugloss	V–IX	English or narrowleaf plantain	V–IX
Common bean	VI–IX	European columbine	VI–VII
Common bird's-foot trefoil	V–IX	European field pansy	V–X
Common butterwort	V–VIII	European hepatica	III–IV
Common chickweed		European speedwell	V–VIII
(*Stellaria media*)	III–X	European yellow-rattle	V–IX
Common evening		Fairy flax	VI–VIII
primrose	VI–IX	Field bindweed	V–X
Common figwort	VI–VII	Field pennycress	IV–VI
Common fumitory	IV–X	Field scabious	V–IX
Common glasswort	VIII–X	Field sow thistle	VII–X
Common groundsel	III–X	Forget-me-not	
Common hemp-nettle	VII–X	Woodland	V–VII
Common hogweed	VI–X	Frostweed	V–IX
Common mare's tail	V–VIII	Garden loosestrife	VI–VIII
Common milkwort	V–VIII	Garden orach(e)	VII–X
Common yarrow	VI–X	Garlic mustard	IV–VI
Corn cockle	VI–IX	Gentian	
Corn poppy	V–VII	Yellow	VI–VIII
Corn spurry	VI–X	Marsh	VII–X
Cornflower	VI–IX	*Gentiana sino ornata*	IV–VIII
Corydalis	III–V	German chamomile	V–IX
Cowslip	IV–V	Germander speedwell	V–VIII
Cow-wheat, small	V–VIII	Globe flower	V–VI
Cranesbill		Golden saxifrage	IV–VI
Bloody	V–IX	Good King Henry	IV–X
Meadow	VI–VIII	Greater celandine	V–IX
Creeping Jenny	V–VII	Ground ivy	IV–VI
Cuckoo flower	IV–VII	Harebell	VI–IX
Cucumber	VI–IX	Hedge false bindweed	VI–IX
Curly dock	V–VII	Hedge mustard	V–X

Species	Months in bloom	Species	Months in bloom
Hellebore, stinking	III–IV	Potato	VI–IX
Hemp agrimony	VII–IX	Purple foxglove	VI–VII
Henbane	VI–X	Purple loosestrife	VI–IX
Hens and chicks	VII–IX	Pussytoes	V–VII
Herb Robert	V–X	Ragged robin	V–VIII
Hoary alyssum	VI–X	Rampion, spiked	V–VIII
Hoary cress	V–VII	Red clover	IV–IX
Honesty	V–VI	Rosebay willow herb	VI–VIII
Hops	V	Sainfoin	V–VII
Hound's tongue	V–VI	Sea holly	
Kidney vetch	V–IX	(*Eryngium maritimum*)	VI–X
Knotweed	V–IX	Sedum	VI–IX
Lady's mantle	V–IX	Self-heal	VI–IX
Large pink	VI–IX	Shepherd's purse	II–IX
Lesser celandine	III–V	Silverweed	V–VIII
Livelong saxifrage	V–VIII	Small scorpion vetch	VI–VII
Lungwort	III–V	Soapwort	VI–IX
Lupin	VI–IX	Solid-stem	
Malve	V–IX	burnet saxifrage	VI–IX
Marsh angelica	VII–VIII	Sorrel	V–VI
Marsh cinquefoil	VI–VII	Spinach	V–IX
Marsh marigold	III–VI	Spring cinquefoil	III–V
Masterwort	VI–VIII	St. John's wort	VI–VIII
Meadow sage	V–VIII	Stinging nettle	VI–X
Meadowsweet	VI–VIII	Strawflower	VII–X
Mercurialis	IV–V	Sundew	VI–VIII
Monkshood	VII–VIII	Sunflower	VII–X
Moschatel	III–V	Swallow-wort	V–VIII
Mouse-ear chickweed	III–VI	Thyme	V–X
Mouse-ear hawkweed	V–X	Tomato	VII–X
Mustard		Tormentil	V–VII
Black	VI–IX	Touch-me-not	VII–IX
Wild	VI–IX	Tufted vetch	VI–VII
Oxeye daisy	V–X	Valerian	VII–VIII
Oxlip	III–V	Vetchling	
Parsley	VI–VII	Spring	IV–V
Pasque flower	III–V	Meadow	VI–VII
Pea	V–VI	Viper's bugloss	VI–X
Peppermint	VI–VII	Water avens	IV–V
Pestilence wort	III–V	Water chestnut	VI–IX
Pincushion flower	VII–X	Water crowfoot	VI–VIII
Pinesap	VI–VIII	Water hemlock	VII–IX
Poison hemlock	VI–IX	Water lily	VI–IX
Pond-lily, yellow	IV–IX	Water parsnip	VI–VIII

(Cont.) →

Species	Months in bloom	Species	Months in bloom
Watercress	V–VIII	Woodruff	IV–V
Whitlow grass	II–V	Woodsorrel	IV–V
Wild chervil	IV–VIII	Wormseed mustard/	
Wild radish	VI–VIII	wallflower	V–IX
Wild strawberry	V–VI	Yellow archangel	IV–VII
Winter cress	IV–VII	Yellow or common toadflax	VI–IX
Wood anemone	III–IV	Yellow sweet clover	V–IX
Wood betony	VI–VIII	Yellowcress, creeping	VI–IX
Wood sanicle	V–VII		

Table 2.4.9: Time of Spore Maturation in Selected Ferns Native to Germany

Data after FLINDT 1985, SCHMEIL and FITSCHEN 1968

Species	Month
Adder's-tongue	VI–VIII
Brackenfern	VII–X
Common clubmoss	VII–VIII
Common moonwort	V–VII
Common polypody	VI–IX
Deer fern	VII–IX
Field horsetail	III–IV
Ground cedar	VIII–IX
Hart's tongue fern	VII–IX
Lady fern	VII–IX
Long beech fern	VI–VIII
Maidenhair spleenwort	VII–IX
Male fern	VI–IX
Marsh clubmoss	VIII–X
Marsh horsetail	V–VII
Meadow horsetail	V–VI
Oak fern	VII–VIII
Royal fern	VI–VII
Spikemoss, Swiss (*Selaginella helvetica*)	VI–VIII
Stiff clubmoss	VIII–X
Wall-rue	VI–VII
Woodland horsetail	V–VI

Table 2.4.10: Formation Time of Fruiting Bodies of Fungi Native to Germany

Data after MAUCH and LAUBER 1975, HENNIG and KREISEL 1978

Species	Months of appearance
Bay bolete	VI–XI
Beefsteak or brain mushroom	III–V
Black morel	III–V
Blusher (*Amanita rubescens*)	VI–X
Brick tuft (*Hypholoma lateritium*)	VIII–XI
Brown roll-rim	VII–XI
Cauliflower mushroom	VII–XI
Chanterelle	VI–X
Common earthball	VII–XI
Death cap	VII–X
Field mushroom	VI–X
Fly agaric	VII–X
Gypsy	VII–X
Honey fungus	IX–XI
Horn of plenty	VIII–XI
Horse mushroom	VI–X
Hygrophorus marzuolus	III–IV
Larch boletus	VII–X
Livid pinkgill (*Entoloma sinuatum*)	V–IX
Orange oak bolete (*Leccinum quercinum*)	VI–X
Oyster mushroom	XI–III
Panther (cap) (*Amanita pantherina*)	VII–X
Parasol	VII–X
Peppery milk cap (*Lactarius piperatus*)	VII–X
Porcini	VI–X
Ramaria aurea (coral fungus)	VIII–X
Red cracking bolete (*Xerocomus chrysenteron*)	VI–XI
Red-staining inocybe (*Inocybe erubescens*)	V–VIII
Saffron milk cap	VII–XI
Sarcodon imbricatus	VIII–XI
Scaly wood mushroom	VI–X
Shaggy ink cap (*Coprinus comatus*)	IV–XI
Sheathed woodtuft (*Kuehneromyces mutabilis*)	IV–XII
Sickener (*Russula emetica*)	VI–X
Stinkhorn (*Phallus impudicus*)	V–IX
Suede bolete (*Xerocomus subtomentosus*)	VI–X
Velvet bolete (*Suillus variegatus*)	VI–XI
White or elfin saddles (*Helvella crispa*)	VIII–XI
White truffle	VII–X
Wood blewit (*Lepista nuda*)	IX–XI
Wood hedgehog (*Hydnum repandum*)	VII–XI
Yellow morel (*Morchella esculenta*)	IV–V

Table 2.4.11: Annual Number of Seeds of Selected Plants

Presented in ascending order, according to number of seeds per plant.

After BERTELSMANN 1979, HANF (no date), MEYER 1964, MÜLLER-SCHNEIDER 1977

Strict or true forget-me-not	56
Coconut palm	70
Wild radish	160
Red dead-nettle	200
Corn cockle	200
Herb Robert	300
Black nightshade	500
Field bindweed	550
Field forget-me-not	700
Common alkanet or bugloss	900
Field pennycress	900
Corn buttercup	900
Wild mustard	1,200
Cornflower	1,600
Pansy	2,500
Hedge mustard	2,700
Creeping bellflower	3,000
Wild carrot	4,000
Dandelion	5,000
Imperforate St. John's wort	10,600
Common chickweed	15,000
Canada thistle	19,000
Orach(e)	20,000
Corn poppy	20,000
Meadowsweet	34,000
Greater celandine	36,000
Groundsel	40,000
Henbane	45,000
Coltsfoot	60,000
Shepherd's purse	64,000
Common spotted orchid	100,000
Horseweed	120,000
Scentless mayweed	210,000
Galinsoga	300,000
Tobacco	360,000
Mugwort	700,000
Butterfly bush	20,000,000
Black poplar	28,000,000
Birch	30,000,000

Table 2.4.12: Number of Seeds per Capsule in Selected Plants

Presented in ascending order, according to number. For the total number of seeds per year, see Table 2.4.11.

Data after BECHTEL et al.1980, EBEL and BIRNBAUM 1971, MEYER 1964

Species	Seeds per capsule (n)
Cherry	1
Coconut	1
Almond	1
Plum	1
Walnut	1
Peanut	3– 4
Apple	5– 10
Pear	5– 10
Pea	8– 12
Shepherd's purse	30– 40
Henbane	300–400
Poppy	>1,000
Maxillaria (orchid)	1,756,440
Swan orchid (*Cycnoches*)	3,751,000 (!)

Table 2.4.13: Weight of Selected Seeds

For the thousand seed weight and the hectoliter weight of the seeds of useful plants, see Table 2.4.14.
Presented in ascending order, according to weight

Data from BERTELSMANN 1979, GÖTZ and KONRAD 1977, MEYER 1964, ULBRICH 1928

Species	Weight (g)
Broomrape (*Orobanche ionantha*)	0.000001
Plantain orchid (*Goodyera repens*)	0.000002
One-flowered wintergreen or pyrola	0.000004
Fragrant orchid	0.000008
Birch	0.0006
Clover	0.002
Alder	0.002
Rape	0.004
Spruce	0.007
Elm	0.008
Douglas fir	0.01
Robinia	0.02
Rye	0.03
European hornbeam	0.035

(Cont.) →

Species	Weight (g)
Wheat	0.04
Oats	0.04
Fir	0.04
Barley	0.04
Ash	0.07
Lupin	0.08
Lentil	0.082
Maple	0.1
Pea	0.2
Corn (maize)	0.2
European beech	0.22
Broad bean	0.3
Common hogweed	0.4
Runner bean (scarlet runner)	0.5
Oak	3.0
Brazil nut	up to 25
Avocado	up to 60

Table 2.4.14: Thousand Seed Weights (TSW) und Hectoliter Weights of the Seeds of Selected Useful Plants

For seed weights, see also Table 2.4.13.

After BROUWER 1977, GÖTZ and KONRAD 1977

Species	TSW (g)	Hectoliter weight (kg)
Winter wheat	35– 45	74–76
Winter rye	25– 35	70–73
Spring barley	35– 45	63–63
Oats	32– 40	50–60
Corn seed	200–450	
Flint corn	115–300	
Sweet corn	240–310	
Popcorn	80–130	
Victoria peas	200–300	78–82
Field peas	150–250	75–82
Broad beans	250–350	75–85
Common vetches	40– 55	76–80
Lupin	100–120	76–80
Rape	4– 6	65–72
Clover	2– 3	70–80
Alfalfa	2– 3	70–80

Table 2.4.15: Age of Seeds of Selected Plants

Age of seeds is in part heavily dependent on storage conditions. The high numbers refer to individual seeds found under the datable foundations of buildings (e.g., churches). For those figures marked with an asterisk (*), there is no information whether the seeds found were still germinable.

Figures from Altman and Dittmer 1972, Gessner 1942, Meyer 1964

Plant species	Max. age of seeds		Plant species	Max. age of seeds	
Sweet chestnut	9	months	Onion	22	years
Horse chestnut	15	months	Melon	30	years
Lemon	16	months	Carrot	31	years
Coconut	16	months	Pea	31	years
Coffee	22	months	Barley	32	years
Beech	2	years	Rye	32	years
Hazelnut	2	years	Wheat	32	years
Common aspen	2	years	Corn (maize)	37	years
Oak	3	years	Celery	39	years
Walnut	5	years	Tobacco	39	years
Eastern white pine	10	years	Oats	41	years
Cornflower	10	years	Mustard	50	years
Pumpkin	10	years	Purple foxglove*	68	years
Poppy	10	years	Dandelion	68	years
Quackgrass	10	years	Common evening primrose*	80	years
Rape	16	years	Red clover	100	years
Millet	17	years	Potato	200	years
Flax	18	years	Sacred lotus	250	years
Cabbage	19	years	Creeping buttercup*	600	years
Hemp	19	years	White clover*	600	years
Lettuce	20	years	Corn spurry*	1,700	years
Bean	22	years			

Table 2.4.16: Maximum Distance Seeds Catapulted by Ballistic Fruits

Listed in ascending order.

Data after Bünning 1953, Kerner v. Marilaun 1898, Strasburger 1983, Ulbrich 1928

Species	Distance catapulted (m)
Narrowleaf bittercress	0.90
Dog violet	1.00
Leopard's bane (*Doronicum decumbens*)	1.00
Longstalk cranesbill	1.50
Marsh cranesbill	2.50
Early dog violet	3.75
Tall violet	4.65
Pansy	4.75
Lupin (*Lupinus digitatus*)	7.00
Bear's breeches (*Acanthus mollis*)	9.50
Squirting cucumber (*Ecballium*)	12.70
Sandbox tree (*Hura crepitans*)	14.00
Bauhinia purpurea (tropical liana)	15.00
As a point of comparison:	
Pilobolus fungus (ejects sporangium)	2.50

2.5 Constituents

Table 2.5.1: Composition of Cereal Grains

Figures are percentages for grains without shells/husks.

After BERTELSMANN 1979, FRANKE 1976, *rororo* 1976/77

	Water	Protein	Fat	Carbo-hydrates	Ash
Wheat	13.2	12.4	2.0	66.7	1.8
Rye	13.7	9.2	1.8	70.2	1.8
Oats	13.0	13.0	7.0	61.3	2.3
Barley	11.7	7.5	1.2	73.6	1.3
Rice	13.1	7.4	0.4	75.6	0.8
Corn (maize)	12.5	11.0	4.4	67.2	1.5
Millet	12.1	11.5	3.9	68.1	1.7

Table 2.5.2: Composition of Legumes

Figures represent percentages.
From FRANKE 1976

	Water	Protein	Fat	Carbo-hydrates	Crude fiber	Minerals
Pea, ripe seeds	10.9	22.90	1.40	60.7	1.40	2.70
Green peas	76.0	6.70	0.50	13.9	1.90	1.00
Bean, ripe seeds	11.6	21.30	1.60	57.6	4.00	3.90
Bean, green beans	90.4	2.24	0.26	5.0	1.36	0.74
Lentil	12.4	25.10	0.70	59.0	0.70	2.10
Broad bean	14.0	23.00	2.00	55.0	6.20	3.10
Soy bean	8.3	40.00	18.80	24.0	4.00	4.60
Peanut, roasted	2.8	26.50	46.60	19.0	2.40	2.70

Table 2.5.3: Fat Content of Selected Plant Organs

Figures represent a percentage of dry weight. See also Table 2.5.4, which provides fat content as a percentage of fresh weight.

Data from FRANKE 1976, STEWARD 1966

Plant species and organ	Fat content (% of dry weight)
Seeds	
Alfalfa	8–10
Almond	40–45
Bean	8
Beechnut	40–46
Cacao	50–58
Castor seed	35–57
Coconut. endosperm	65–72
Coffee	5–10
Corn (maize)	4.7
Cotton	16–24
Flax	24–43
Hazelnut	60–68
Hemp	30–35
Horse chestnut	3– 6
Oil palm	50
Parsley	16–19
Peanut	24–56
Pecan	65
Poppy	40–51
Rape	22–49
Rice	2.2
Sesame	44–54
Soy bean	17–22
Sunflower	40–65
Walnut	64
Wheat	2– 2.4
Fruit pericarp	
Avocado	70
Bay laurel	24–55
Oil palm	71–81
Olive	35–70
Leaves/Blades	
Alfalfa	2.3
Cabbage	1.7
Orchard grass	2.2

Plant species and organ	Fat content (% of dry weight)
Rye-grass (Lolium perenne)	1.7
Spinach	0.4
Roots	
Beet (*Beta vulgaris*)	7
Mangrove	70
Sprouts	
Black-eyed pea (*Vigna sinensis*)	2.6
Soy bean	2.5
Tilia (linden, lime) (*T. americana*). rind	2.3

Table 2.5.4: Fat Content of the Seeds and Fruits of Selected Plants as a Percentage of Fresh Weight

Compare also to Table 2.5.3, which provides the fat content of the dry weight of selected plant organs.

Data from MENGEL 1968

Species	Fat content (% of fresh weight)
Barley	2
Corn (maize)	4.8
Flax	40
Oats	5
Olive	20
Olive (fruit flesh)	53
Pea	2
Peanut	45
Poppy	45
Rape	42
Soy bean	36
Sunflower (shelled)	50
Wheat	1.7

Table 2.5.5: Raw Protein Content of Selected Seeds and Fruits

Listed in ascending order.

From MENGEL 1968

Species	(%)
Rice	6.9
Corn (maize)	9.6
Barley	9.7
Oats	10.7
Millet	12.7
Flax	22.5
Pea	23.2
Broad bean	25.3
Sweet pea	25.9
Yellow lupine	38.3

Table 2.5.6: Composition of Tubers

Figures represent the mean values of amount present in an edible portion of 100 g.

From SOUCI, FACHMANN and KRAUT 1982

	Sweet potato		Cassava		Yams		Potatoes	
Water	69.20	g	65.00	g	72.00	g	77.80	g
Protein	1.63	g	0.90	g	1.79	g	2.00	g
Fat	0.60	g	0.35	g	0.24	g	0.15	g
Carbohydrates	26.60	g	32.00	g	23.77	g	18.90	g
Raw fiber	0.90	g	0.80	g	0.71	g	0.78	g
Minerals	1.12	g	0.40	g	1.00	g	1.10	g
Vitamin B_1	0.064	mg	0.07	mg	–		0.11	mg
Vitamin B_2	0.05	mg	–		–		0.051	mg
Nicotinamide	0.60	mg	–		–		1.22	mg
Vitamin C	17.0	mg	25.0	mg	–		15.0	mg

Table 2.5.7: Sugar Content of the Nectar of Selected Plants

After SPECTOR 1956

Plant	Sugar content (g per 100 cm³ nectar)	Sugar formation per blossom in 24 h (mg)
Alfalfa	33	0.1
Apple	75	0.8
Blackberry	15.4–45	
Borage	22 –29	
Cucumber	30	1.0
Fireweed	18.4–41	
Horse chestnut	69	1.1
Hound's tongue	36	0.3
Ivy	10 –11.5	
Jasmine	33	0.3
Lavender	24 –32	
Nasturtium	46	1.7
Pear	16	0.3
Pumpkin	28	28
Rape	47	0.5
Raspberry	46	7.6
Red clover	8.3–30	
Robinia	55	1.0
Snapdragon	45	1.5
Sour cherry	15	1.2
Sunflower	38	0.3
Tilia (linden, lime)	21 –42	
White dead-nettle	46	1.7
White clover	25 –39	
White mustard	9.3–17.6	
White sweet clover	36	0.04
Wild cherry	21	0.5

Table 2.5.8: Energy Contents of Plant Material

Expressed in kcal (kJ) per g dry weight. Figures are mean values

From MÜHLENBERG 1976, modified

Leaves	4.2 kcal (17,6 kJ)
Stalks and stems	4.3 kcal (18.0 kJ)
Roots	4.7 kcal (19.7 kJ)
Seeds	5.1 kcal (21.3 kJ)
Straw	4.3 kcal (18.0 kJ)

2.6 Physiological Values

Table 2.6.1: Speed of Plasma Flow in Selected Plant Organs

See also Table 5.6.

After OPPENHEIMER and PINCUSSEN 1927

Plant	Cell	Speed (μm/s)
Arrowhead	Leaf	2.9
Canadian waterweed	Leaf	15.4
Frog's bit	Root hair	9.5
Henbane	Hair	12
Mouse-ear chickweed	Leaf	1.5
Mucor mold	Hyphae	55
Nettle	Hair	6
Nitellaflexilis	Internode	27.1
Onion	Onion scale epidermis	7 – 8
Phycomyces nitens	Hyphae	55
Pumpkin	Hair	8.3–10
Tradescantia	Stamen hair	10.8–13.8
Vallisneria spiralis	Leaf	12 –26

Table 2.6.2: Osmotic Pressure in the Cell Sap of Selected Plant Leaves

For figures for entire plants, see Table 2.6.3.

After ALTMAN and DITTMER 1966, SPECTOR 1956

Plant	Osmotic pressure atm	(hPa)	Plant	Osmotic pressure atm	(hPa)
Acacia	18–35	(18,239–35,464)	Common		
Alfalfa	16.3–24.1	(16,516–24,419)	glasswort	30.9–113	(31,309–114,497)
Apple	24–27	(24,318–27,358)	Corn (maize)	5.7–16.2	(5,776–16,415)
Asarabacca (European wild ginger)			English holly	15.5–19.5	(15,705–19,758)
	9.0	(9,119)	European beach or marram grass		
Ash	15–22.5	(15,199–22,798)		13.8–40.0	(13,983–40,530)
Barley	7.2	(7,295)	European hornbeam		
Birch	12–23	(12,159–23,305)		17.7–24.3	(17,935–24,622)
Canadian waterweed			Fig	14.6	(14,793)
	6.7	(6,789)	Fir	17.1–22.3	(17,327–22,595)
Carrot	13.2–16.7	(13,375–16,921)	Fodder beets	11.3	(11,450)
Cattail	18.1–20.8	(18,340–21,076)	Gypsophila	15–70	(15,199–70,928)

Plant	Osmotic pressure atm	(hPa)	Plant	Osmotic pressure atm	(hPa)
Hellebore	23.8	(24,115)	Poplar	17.8	(18,036)
Holm oak	9.8–20.1	(9,930–20,366)	Potato	5.2–9.3	(5,269–9,423)
Ivy	12–25	(12,159–25,331)	Pumpkin	7.5–9.8	(7,599–9,930)
Larch	13.3–21.3	(13,476–21,582)	Rubber tree	7.9–9.9	(8,005–10,031)
Mangrove	30	(30,398)	Sod (*Suaeda*)	130	(131,723)
Mistletoe	25 –30	(25,331–30,398)	Sunflower	8.6–16.4	(8,714–16,617)
Oats	9.4–15.8	(9,525–16,009)	Tobacco	7.7–19.0	(7,802–19,252)
Onion	9.2–11.6	(9,322–11,754)	Tomato	9.5–14.4	(9,626–14,591)
Peach	25 –35	(25,331–35,464)	Wheat	9.6–28.0	(9,727–28,371)
Pine	16.1	(16,313)	Wine grape	9.3–23.1	(9,423–23,406)

Table 2.6.3: Osmotic Values in Plants

Unless otherwise indicated, the figure given holds for the plant as a whole. For leaf data, see Table 2.6.2. Listed in ascending order.

After ALTMAN and DITTMER 1966, GESSNER 1942, MADLE 1967, WALTER 1962

Plant	Osmotic pressure atm	(hPa)
Hydrophytes	1 – 3	(1,013–3,040)
Succulents	5 – 7	(5,066–7,093)
Potato, tuber	7	(7,093)
Watermilfoil	8	(8,106)
Tomato, fruit	9	(9,119)
Helophytes (marsh plants)	9 –14	(9,119–14,186)
Potato, shoot	10	(10,133)
Cereal grains	10 –17	(10,133–17,225)
Pondweed	11	(11,146)
Alga (*Spirogyra*)	11.1–12.4	(11,247–12,564)
Canadian waterweed	11.5	(11,652)
Lemon	14	(14,186)
Sugar beet, root	14 –21	(14,186–21,278)
Conifer, needles	16 –22	(16,212–22,292)
Apple, fruit	18 –23	(18,239–23,305)
Desert xerophytes	20 –40	(20,265–40,530)
Lilac, upper epidermis	23.4	(23,710)
Red alga (*Ceramium*)	35 –37	(35,464–37,490)
Sea lettuce (Alga)	37 –46	(37,490–46,610)

(Cont.) →

Plant	Osmotic pressure atm	(hPa)
Wine grape, fruit	38	(38,504)
Sargassum (brown algae)	39– 42	(39,517–42,557)
Sugar kelp (*Laminaria*)	50– 55	(50,663–55,729)
Palmleaf cinquefoil	73.8	(74,778)
Wall germander	83	(84,100)
Halophytes	>100	(>101,325)

Table 2.6.4: Acidity in the Pressed Juice of Fruits and Vegetables

After SPECTOR 1956, WALTER 1962

Species	Acidity level	Species	Acidity level
Apple	pH 3.4	Orange. ripe	pH 3.8
Blackberry	pH 2.7	Orange. unripe	pH 2.9
Celery	pH 5.2	Pineapple	pH 4.4
Cowberry (lingonberry)	pH 2.4	Potato	pH 5.6
Grapefruit	pH 3.0	Rhubarb	pH 3.2
Lemon	pH 2.4	Tomato	pH 4.4
Onion	pH 4.3	Turnips	pH 5.8

Table 2.6.5: Temperature Tolerances of Plants

Temperature tolerances vary with adaptation temperature. Specimens from colder regions have a greater tolerance for low temperatures and a lesser tolerance for high ones. The reverse holds for specimens from warm regions. Values are also strongly dependent upon the water content of the given plant and upon the plant organ studied.

Data after various authors in ALTMAN and DITTMER 1973

Species	Minimum (°C)	Maximum (°C)
Algae		
Bladder wrack	− 20	41.9
Chlamydomonas nivalis	− 36	4
Marine red algae	− 2 up to + 3	32–35
Oscillatoria filiformis	59	83
Sea lettuce (*Ulva*)	− 8	30

Species	Minimum (°C)	Maximum (°C)
Spirogyra	− 7	40
Sugar kelp	− 4	24

Lichens

Lung lichen ("lungwort")	− 196	73
Reindeer lichen	− 196	

Mosses

Bryum sp.	− 7	42
Frullania dilatata		70–75
Plagiothecium curvifolium		60–75

Ferns

Brackenfern (rhizome)	− 2.5	
Common polypody	− 18.1	48.5
Hart's tongue fern	− 14.8	47.5

Spermatophytes (seed-bearing plants)

Alder, buds	− 28	
Alder, leaves	− 3	
Ash, buds	− 27	
Ash, leaves	− 2	
Barley	− 15 up to − 10	
Canadian waterweed		39–39.5
Cross-leaved heath	− 20	50.5
European beech, buds	− 22	
European beech, leaves	− 2.5	
Field or hedge maple, buds	− 23	
Field or hedge maple, leaves	− 2.5	
Oats	− 12 up to − 9	
Rye	− 25 up to − 15	
Silver fir (summer)	− 4	46
Spruce (summer)	− 8 up to − 3.5	42.5
Spruce (winter)	− 38	40.5
Swiss pine (summer)	− 10	
Swiss pine (winter)	− 40	
Wheat	− 22 up to − 10	
Woodsorrel	− 11.5	40.5
Yew (summer)	− 4	50.5
Yew (winter)	− 40	47.5

Table 2.6.6: Annual Levels of Evaporation of Selected Plants

For further figures on the evaporation levels of plants, see Tables 2.6.7 and 2.6.8. Figures represent averages.

After MADLE 1967

Plant	Annual evaporation level	
1 corn (maize) plant	100–180	liters
1 ha grain	1.5–2.5	million liters
1 ha beech forest	3.6	million liters
1 ha spruce forest	3.2–3.9	million liters
1 ha pine forest	1.6–2.7	million liters

Table 2.6.7: Transpiration Levels of Selected Plants

Transpiration values (cuticular + stomatal) in milligrams water per 100 cm^2 leave surface area and hour. For figures on cuticular transpiration, see Table 2.6.8. Evaporation is dependent upon many external factors, such as temperature (T), soil moisture (SM), relative air humidity (RH), light, and wind levels. In so far as these values are known, they are provided here under "External conditions."

Figures from ALTMAN and DITTMER 1974

Species	Water loss (mg per 100 cm^2/h)	External conditions
Banana	1,100	T = 25 °C, RH = 32 %
Barley	958	SM = 15 %
Bean	450	T = 30 °C, RH = 35 %
Millet	42	T = 30 °C
Oats	1,542	SM = 15 %
Pineapple	140	T = 25 °C
Sunflower	576	T = 30 °C, RH = 35 %
Wheat	84	T = 30 °C
Wine grape	473	T = 20 °C

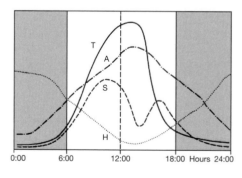

0:00   6:00   12:00   18:00 Hours 24:00

Fig. 2.6.7: Daily course of transpiration (schematic depiction) of a plant. See also Table 2.6.7. T = transpiration, A = air temperature, S = openness of stomata, H = humidity. From SCHMEIL and SEYBOLD 1958

Table 2.6.8: Cuticular Transpiration in Selected Plants

Values obtained under standard conditions. Figures represent milligrams water per hour and per gram fresh weight. Listed in descending order of transpiration.

Figures from PISEK and BERGER 1938

Species	Cuticular transpiration (mg per h/g fresh weight)
Touch-me-not	130
Lungwort	71
Marsh marigold	56.5
Wood stitchwort (*Stellaria nemorum*)	56
European speedwell	55
Hazel	47
Yellow woundwort	33.5
Crown vetch	33
Field bindweed	25
European white birch	25
European beech	25
Oak	24.5
Asarabacca (European wild ginger)	24
Alpine rose. rust-colored	7.25
Witch's moneybags (*sedum telephium*)	5.1
Bearberry	5.0
Ivy	3.1
Spruce	1.58
Pine	1.53
Prickly pear cactus	0.12

Table 2.6.9: Water Consumption for Selected Plants Required for Production of 1 Kg Plant Dry Matter (So-Called Transpiration Coefficient)

For figures on evaporation, see also Tables 2.6.6 - 2.6.8.

After BERTELSMANN 1979, MADLE 1967, MENGEL 1968, WALTER 1962

Species	(liters)	Species	(liters)
Alfalfa	900	Red clover	700–800
Barley	530	Rye	400–700
Bean	625	Spruce	300
Beech	400	Sugar beet	400–440
Corn (maize)	300–400	Sunflower	570
Flax	780	Vegetables	800
Oats	400–600	Wheat	450–600
Pine	160	Wild mustard	910
Potato	300–600		

Table 2.6.10: Hydraulic Conductivity of Selected Woods

Figures represent the amount of water that can flow, per square centimeter, through a section of sprout under controlled experimental conditions (15-cm-long section, pressure of 30 mmHg). Since the experimental conditions do not correspond to natural ones, the resulting figures can only be compared in a relative sense. Presented in ascending order, according to amount. On vessel sizes, see also Table 2.2.2.

After FARNER 1918

Species	Hydraulic conductivity	Species	Hydraulic conductivity
Ruscus	1.0	Field or hedge maple	45.0
Mezereon	6.5	Apple	45.0
Tree heath	8.0	European spindle	47.0
Cherry laurel	10.0	Birch	51.0
Yew	12.0	Ivy	60.0
Larch	14.5	European beech	65.0
Scotch broom	25.0	Oak	75.0
Hazelnut	31.0	Eurasian aspen	81.0
Privet	37.0	Basket willow	95.0
Hawthorn	44.0		

Table 2.6.11: Speed of Hydraulic Conductance in Selected Plants

Speed is dependent upon the transverse diameter of the water vessels; here, see Table 2.2.2.

After GESSNER 1942, KAUSSMANN 1963, KREMER 1980, MEYER 1964, ZIEGLER 1977

Species	m/h	Species	m/h
Algae (*Macrocystis*)	0.65–0.78	Willow	3.00
Horse chestnut	0.80	Tilia (linden, lime)	3.43
European beech	1.07	Walnut	4.12
European hornbeam	1.25	Elm	6.00
Conifers	1.50	Ash	25.70
Birch	1.60	Robinia	28.80
Alder	2.00	Oak	43.60
Maple	2.40	Lianas	150.00
Olive	2.40		

Table 2.6.12: Respiration Rates of Selected Lower Plants (Thallophyta)

Figures are in cubic millimeters of O_2 per gram and per hour of either dry or fresh weight. For the respiration rates of seed-bearing plants (spermatophytes), see Table 2.6.13; for that of bacteria and fungi, see Table 3.6.

Figures from ALTMAN and DITTMER 1974

Species	O_2 consumption/h and	
	per 1 g fresh weight	per 1 g dry weight
Algae		
Bladder wrack	200	
Cladophora	330	
Red alga (*Delesseria*)		41
Sargassum		24
Sea lettuce (*Ulva*)		810
Toothed wrack (*Fucus serratus*)	190	
Mosses		
Hypnum cupressiforme	300	
Peat moss		71
Riccia fluitans	3,000	
Lichens		
Iceland moss		48
Lobaria linita		72
Parmelia acetabulum		31
Reindeer lichen		85

Table 2.6.13: Respiration Rates of Selected Plants and Plant Parts

Figures are in cubic millimeters of O_2 per gram and per hour of either dry or fresh weight. Figures are not always comparable to one another, since there is a strong dependence of the respiration rate upon the temperature, moisture content, and general condition of the plant. For figures on lower plants (thallophyta), see Table 2.6.12; for bacteria and fungi, see Table 3.6.

Data after ALTMAN and DITTMER 1974, STEWARD 1960, STILES and COCKING 1974

Species and organ	O_2 consumption/h and	
	Per 1 g fresh weight	Per 1 g dry weight
Apple, fruit	20– 35	
Barley, leaves	266	
Barley, seeds	0.06	0.05
Bean, entire plant	96.6	
Bean, leaf	260– 570	

(Cont.) →

Species and organ	O$_2$ consumption/h and	
	Per 1 g fresh weight	Per 1 g dry weight
Beet (*Beta vulgaris*), root	9	
Canadian waterweed, shoot	900	
Carrot, root	25–30	
Carrot, young leaves	1,133	
Corn (maize), seeds		2.4
Cuckoopint, shoot	15,600–31,800	
Fir, shoot	250	
Flax, seeds		0.3
Lily, blossom		12,670
Oats, seeds		0.2
Onion, root	11,400	
Pea, leaf		4,300
Pineapple, fruit	30	
Potato, tuber	33	
Rye, seeds		0.02
Snowdrop, entire plant	77.6	
Spinach, leaf	620	
Spruce, entire plant	44.1	
Timothy grass, blade		1,240
Tobacco, entire plant	77	
Tobacco, leaf		3,300
Tomato, leaf		3,900
Tulip, blossom		3,000
Wheat, entire plant	291	
Wheat, seeds		0.05

Table 2.6.14: Sugar and Starch Production in Photosynthesis

Figures are per 1 m^2 leaf surface and per hour under optimal conditions

From MEYER 1964

Species	Result of photosynthesis	
	Sugar (g)	Starch (g)
Autumn crocus	1.217	
Cuckoopint	1.004	
Dock		2.215
Lauch	1.193	
Sunflower		1.823
Tobacco		1.378
Tulip	1.267	
Water lily		2.273

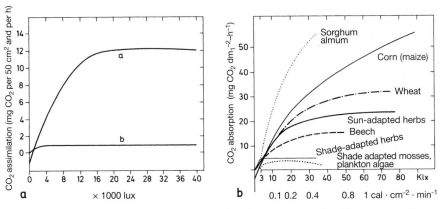

Fig. 2.6.14: Dependence of results of photosynthesis on leaves upon level of illumination. **a** Here *a* represents a white mustard sun leaf, and *b* represents a woodsorrel shade leaf. See also Table 2.6.14. From GUTTENBERG 1963, p. 503. **b** Dependence of net photosynthesis upon exposure to light in selected plants. From LARCHER 1976, p. 62

Table 2.6.15: Annual Worldwide Photosynthesis Production and Carbon Exchange

Figures from JENSEN and SALISBURY 1972, LIBBERT 1979

	Surface area (10^6 km^2)	Net annual primary production (10^{12} kg)	Annual production of organic carbon (10^{12} kg)	Annual consumption of CO_2 (10^{12} kg)	Annual energy storage (10^{12} kJ)
Oceans	361	55.00	25.0	92	1.0
Land, total	149	108.00	50.0	183	2.0
Rivers and lakes	2	1.00			
Swamps, moors Marshes	2	4.00			
Tropical forests	20	40.00			
Forests, mixed zones	18	23.40			
Taiga	12	9.60			
Brushland (macchias, dwarf pine region, and similar)	7	4.20			
Tropical savannahs	15	10.50			
Grasslands, mixed zones	9	4.50			
Tundra	8	1.10			
Deserts	18	1.30			

(Cont.) →

	Surface area (10^6 km^2)	Net annual primary production (10^{12} kg)	Annual production of organic carbon (10^{12} kg)	Annual consumption of CO$_2$ (10^{12} kg)	Annual energy storage (10^{12} kJ)
Rocks, ice	24	0.07			
Area under cultivation	14	9.10			
Earth as a whole	510	165	75	275	3.0

Table 2.6.16: Speed of Rotation in Selected Twining Plants

Figures represent the time it takes for a twining plant to complete a circle while making searching movements. Listed in descending order, according to time taken.

Data after KERNER V. MARILAUN 1896, STILES and COCKING 1969

Species	Hours per round
Roxburghia viridifolia	24
Bellflower *(Lapageria rosea)*	14
Jasmine *(Jasminum pauciflorum)*	7
Fern *(Lygodium scandens)*	6
Glycine *(Wistaria sinensis)*	3
Ruscus androgynus	3
Bryony *(Tamus communis)*	3
Hops	2
Runner bean (scarlet runner)	2
Hedge false bindweed	2
Grape vine	1

Table 2.6.17: Resting Potential of Plant Cells and Concentration of Potassium Ions in the Membrane (Plasmalemma)

For figures on animal cells, see Tables 1.10.5 and 1.10.6.

Data from ALTMAN and DITTMER 1973

Species	Resting potential (mV)	K^+ ion concentration (mmol) Outside cell	Inside cell
Chara corallina	− 140	0,2	120
Enteromorpha intestinalis	− 42	11,0	450
Nitella translucens	− 138	0,1	119
Valonia ventricosa	− 71	11,0	434
Hydrodyction africanum	− 116	0,1	93

Table 2.6.18: Conduction Velocity in Selected Plants

For the corresponding figures in animals, see Table 1.10.7. On the production and generation of excitation in plants, see also STRASBURGER 1983.

Data after GESSNER 1942, STRASBURGER 1983

Species and organ	Conduction velocity	
Nitella, internodal cell	2.3	cm/s
Mimosa, primary petiole	0.4– 3	cm/s
	max. 10	cm/s
Venus flytrap, leaf	6–20	cm/s
Waterwheel plant (*Aldrovanda*), leaf	6–20	cm/s

Table 2.6.19: Absorption Maxima of Plant Pigments

Figures represent wavelengths measured in nanometers. Since absorption is also dependent on the solvent used, it is also provided. The wavelength with the main absorption is underlined.

Data after SPECTOR 1956

Pigment	Absorption maxima (nm)	Solvent
Phycoerythrin	576	Acidic $CHCl_3$
Phycocyanin	630	Acidic $CHCl_3$
Beta-carotene	497, 466	Chloroform
Xanthophyll	487, 456, 428	Chloroform
Flavone	298, 250	Ethanol
Chlorophyll a	<u>660</u>, 617, 575, 430, 410	Ether
Chlorophyll b	<u>640</u>, 595, 535, <u>455</u>	Ether
Chlorophyll c	<u>627</u>, <u>597,5</u>, 545, 446	Ether
Chlorophyll d	<u>686</u>, 645, 595, 545, <u>445</u>, 390	Ether
Protochlorophyll	<u>623</u>, 571, 535, 432	Ether

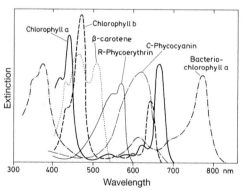

Fig. 2.6.19: Absorption spectra of several important biological pigments. After LIBBERT. From STRASBURGER 1983, p. 233

Part 3: Microbiology

Table 3.1: Size and Structure of Viruses

The term virion is the more recent designation for an infective viral particle; capsomer refers to the morphological subunit that builds the capsule that envelops the nucleic acid. The virus that triggers AIDS is one of the retroviruses. For data on the size of bacteria, see Table 3.2.

Data after GRAFE 1977, LASKIN and LECHEVALIER 1973, OTTE and BRANDIS 1978, RIESNER 1979, WEIDE and AURICH 1979

Virus group	Diameter of virion (nm)		Capsomers (n)	Molecular weight of the genome ($\times 10^6$)
Adenoviruses	70– 90		252	20 – 30
Arenaviruses	50–300			3.2– 5.6
Bacteriophage λ	54			33
Coronaviruses	80–160			9
Herpesviruses	160–200		162	92 –102
Iridoviruses	130			130
Orthomyxoviruses (influenza viruses)	80–120			6
Papovaviruses	45– 55		72	3 – 5
Paramyxoviruses	150–300			5 – 8
Parvoviruses	16– 26		32 ?	1.5– 2.2
Picornaviruses	20– 30		32 ?	2.5
Pox viruses	170–260	× 300–450		130 –240
Reoviruses	60– 80		92 ?	15
Retroviruses	80–120			7 – 10
Rhabdoviruses	70 × 130	–300		3.5– 4.5
Togaviruses	40– 70		32	3 – 4
Viroid	40 ×? 2			0.12

Table 3.2: Dimensions of Bacteria

For data on viruses, see Table 3.1.

Data after KUNSTYR et al. 1988, OTTE and BRANDIS 1978, SCHLEGEL 1976

Species	Length (µm)	Width (µm)
Bacillus anthracis	3.0–10.0	1.0–1.5
Bacillus megatherium	5.0	2.0
Bordetella pertussis	0.5– 1.0	0.3–0.5
Brucella spp.	0.6– 1.5	0.5–0.7
Corynebacterium diphtheriae	0.2	
Clostridium botulinum	2.0–10.0	0.5–1.0
Clostridium tetani	2.4– 5.0	0.5–1.1

(Cont.) →

Species	Length (µm)	Width (µm)
Enterobacteria	2.0– 3.0	0.5–1.0
Francisella tularensis	0.2– 0.7	
Haemophilus influenzae	0.3– 0.5	
Leptospires	7.0– 25.0	0.12
Leptotricha buccalis	5.0– 15.0	1.0–1.5
Metabacterium criceti	20.0	4.0
Micrococcus	0.2	
Mycobacterium tuberculosis	3.0	0.4
Mycoplasmas	0.3– 0.8	
Neisseria gonorrhoeae	0.8	
Pseudomonas spp.	0.5– 1.0	1.5–4.0
Rickettsia	1.0– 2.0	0.3–0.7
Spirochetes	30.0–500.0	
Staphylococcus	0.8– 1.0	
Streptococcus pneumoniae	0.8– 1.0	
Vibrio cholerae	1.5– 3.0	0.5

Table 3.3: Generation Time of Selected Bacteria

Generation time in bacteria represents the time interval from one cell division to the next. Since it is heavily dependent upon temperature, the table also gives the latter. The in part very large ranges result from the fact that generation time can be dependent upon the given culture medium.

Data after ALTMAN and DITTMER 1972, SPECTOR 1956

Bacterial species	Temperature (°C)	Generation time (min)
Bacillus cereus	37	18.8
Bacillus subtilis	36	35
Clostridium botulinum	37	35
Diplococcus pneumoniae, Type I	37	20.5–29
Diplococcus pneumoniae, Type II	37	23.0–33
Escherichia coli	37	12.5–17
Lactobacillus acidophilus	37	60.0–87
Proteus vulgaris	37	21.5–46
Pseudomonas aeruginosa	37	31.0–34
Rhizobium leguminosarum	25	79.0–187
Salmonella typhimurum	37	24.0–35
Salmonella typhosa	37	23.5–33
Serratia marcescens	35	28
Staphylococcus aureus	37	27.0–32
Streptococcus lactis	37	26.0–37
Vibrio cholerae	37	21.2–38

Table 3.4: pH Ranges for the Growth of Bacteria and Fungi

Figures from Thimann 1964

	Optimum	Minimum	Maximum
Bacteria, cocci			
Diplococcus pneumoniae	7.8	7.0	8.3
Neisseria meningitidis	7.4	6.1	7.8
Neisseria gonorrhoeae	7.3	5.8	8.3
Streptococcus pyogenes	7.8	4.5	9.2
Bacteria, rods			
Agrobacterium tumefaciens		5.7	9.2
Azotobacter chroococcum	7.4–7.6	5.8	
Bacillus subtilis	6.0–7.5	4.5	8.5
Clostridium tetanis	7.0–7.6	5.5	8.3
Escherichia coli	6.0–7.0	4.3	9.5
Mycobacterium tuberculosis	6.8–7.7	5.0	8.4
Pasteurella pestis	6.2–7.0	5.0	8.2
Proteus vulgaris	6.5	4.4	9.2
Salmonella parathyphosum	6.2–7.2	4.0	9.6
Salmonella thyphosum	6.8–7.2	4.0	9.6
Serratia marcescens	6.0–7.0	4.6	8.0
Vibrio cholerae	7.0–7.4	5.6	9.6
Fungi			
Fusarium oxysporium		1.8	11.1
Mucor glomerula		3.2	9.2
Penicillium variabile		1.9	9.3
Pilobulus kleinii		6.2	10.1
Saccharomyces cerevisiae	4.0–5.0	2.4	

Table 3.5: Lethal Temperatures for Selected Microorganisms

Temperature values are accompanied by figures for the time required to kill the respective organism. For data on viruses, see Table 3.6.

Figures from Altman and Dittmer 1973, Schlegel 1976

Microorganism	Lethal temperature (°C)	Time required (min)
Bacteria		
Aerobacter aerogenes	55	30
Bacillus anthracis, vegetative form	60	30

(Cont.) →

Microorganism	Lethal temperature (°C)	Time required (min)
Bacillus anthracis, spores	100	5–10
Bacillus subtilis, veg. form	95	36
Bacillus subtilis, spores	100	15–20
Clostridium botulinum, spores	120	5
Clostridium tetani, veg. form	105	3–25
Clostridium tetani, spores	105	3–25
Diplococcus pneumoniae	55	20
Escherichia coli	60	15
Lactobacillus thermophilus	71	30
Mycobacterium tuberculosis	60	15
Neisseria gonorrhoeae	55	<5
Neisseria meningitidis	55	<5
Pseudomonas aeruginosa	55	60
Salmonella typhi	60	4.3
Staphylococcus aureus	60	10
Streptococcus pyogenes	55	30
Vibrio cholerae	55	15
Fungi		
Aspergillus niger, spores	55	30
Candida albicans	60	10
Fusarium sp.	55	10
Monilia fruticola, spores	52	2–3
Mucor sp.	57.5	30
Penicillium sp.	50–60	30
Saccharomyces cerevisiae	68	30

Table 3.6: Inactivation Temperatures and Times for Selected Viruses

For the corresponding figures for bacteria and fungi, see Table 3.5.

Data after GRAFE 1977

Virus	Inactivation temperature (°C)	Inactivation time (min)
Adenoviruses	50	20
Avian flu virus	60	10
Foot-and-mouth disease virus	56	30
Herpes simplex virus	50	30
Influenza virus	56	30

Virus	Inactivation temperature (°C)	Inactivation time (min)
Mumps virus	55	20
Parnoviruses	56	60
Poliomyelitis virus	50	30
Yellow fever virus	60	10

Table 3.7: Respiration Rate of Selected Bacteria and Fungi

Figures given in cubic millimeters O_2 per milligram dry weight and per hour. For data on the respiration rates of lower plants, see Table 2.6.12 and for those of sead-bearing plants, see Table 2.6.13.

Figures from ALTMAN and DITTMER 1974

Species	O_2 consumption (mm³ O_2/mg per h)
Bacteria	
Azotobacter chroococcum	2,000
Bacillus subtilis	170
Escherichia coli	272
Lactobacillus bulgaricus	
Leuconostoc sp.	8
Micrococcus aurantiacus	14
Mycobacterium tuberculosis	4
Neisseria meningitidis	56
Pseudomonas aeruginosa	26
Sarcina lutea	18.3
Streptococcus pneumoniae	27
Streptococcus pyogenes	57–63
Fungi	
Aspergillus niger	2.99
Candida albicans	29
Fusarium graminearum	13.6
Monilia fruticola	3.21
Neurospora crassa	29.6
Penicillium	6.5
Puccinia graminis	3
Saccharomyces cerevisiae	47
Saprolegnia	0.75

Table 3.8: Incubation Times of Several Important Infectious Diseases

Figures from GRAFE 1977, *Lexikon der Biologie*, OTTE and BRANDIS 1978, SIEWING 1985

Disease	Incubation times
AIDS	2 weeks – 6 years
Amoebic dysentery	1 day – 4 weeks
Anthrax	2– 3 days
Chicken pox	11–21 days
Cholera	1– 5 days
Diphtheria	2– 5 days
Dysentery (shigellosis)	2– 7 days
Epidemic hepatitis	1– 7 weeks
German measles (rubella)	14–21 days
Gonorrhea	1– 2 days
Herpes zoster (shingles)	2– 3 weeks
Influenza	1– 3 days
Leprosy	9 months to several years
Malaria (*M. tropica*)	7–12 days
Measles	9–14 days
Mumps	18–21 days
Pertussis (whooping cough)	8–15 days
Poliomyelitis	7–14 days
Rabies	20–60 days
Scarlet fever	2– 7 days
Smallpox	5–14 days
Syphilis	14–21 days
Tetanus	4–14 days
Typhus	5–21 days
Typhus	1– 3 days
Yellow fever	3– 6 days

Part 4: Human Biology

Isolated data pertaining to humans can also be found in most of the tables in Part I, Zoology. More comprehensive data can also be gained from the follow-on volume: KUNSCH and KUNSCH (2001) *Der Mensch in Zahlen*

4.1 General Data

Table 4.1.0: Numbers that Amaze

This table brings together some interesting numbers on humans, which are cited again in the individual tables below, where they are accompanied by references to the literature.

Number of cells	in the skin	110,000,000,000
	per cm²	6,000,000
	in the brain	14,000,000,000
Number of mucous membrane cells	in the small intestine	200,000,000,000
Number of hairs on a blonde person	on the head	150,000
	+ body hair	25,000
Number of perspiratory glands in total	on the skin	1,800,000
	per cm²	100
Number of heart beats	in 70 years	3,000,000,000
Amount of blood conveyed	in 70 years	250,000,000 liters
Length of the blood capillaries	in total	1,200 km
Number of alveoli	in the lung	850,000,000
Length of capillaries	in the alveoli	13 km
Number of salivary glands in total	in the mouth	6,000,000
Number of intestinal villi	in the small intestine	7,000,000
Number of renal corpuscles (glomeruli)		2,000,000
Length of the loops of Henle	in total	100 km
Number of synapses	in the brain	1,000,000,000,000
Length of nerve fibers in total	in the cerebrum	500,000 km
Number of retinal cones	in the eye	7,000,000
Number of retinal rods	in the eye	125,000,000
Number of hair cells on the basilar membrane		
	in the ear	14,000

Table 4.1.1: Life Span of Cells in Selected Organs of the Human Body

Data after KLIMA 1967, RUCKER 1967

Organ	Average life span (days)
Stomach (pylorus)	1.8– 1.9
Stomach (cardia)	9.1
Small intestine	1.3– 1.6
Liver	10.0– 20.0
Large intestine	10.0
Rectum	6.2
Anus	4.3
Trachea	47.5
Lungs (alveoli)	8.1
White blood cells	1.0– 3.0
Red blood cells	120.0
Bladder	64.0
Epidermis	
Lips	14.7
Soles	19.1
Abdominal skin	19.4
Ear	34.5
Nervous system	No regeneration

Table 4.1.2: DNA and RNA Contents of Selected Human Cells

Figures in picograms (1 pg = 1 trillionth of a gram = 0.000 000 001 g) per cell. For comparative figures for animals, see Table 5.11.

Figures from ALTMAN and DITTMER 1972

Cell	DNA (pg/cell)	RNA (pg/cell)
Bone marrow	0.87	0.69
Brain	0.68	2.63
Kidney	0.83	1.10
Liver	1.0	2.48
Leukocytes	0.73	0.25
Sperm	0.31	0.24

Table 4.1.3: Composition of the Human Body from the Elements

Figures refer to a human weighing 70 kg. For comparable figures for plants and animals, see Table 5.1.

After HEIDERMANNS 1957, KLEIBER 1967

Element	As approx. percentage of total weight (%)	Approx. weight	
Oxygen (O)	63	44	kg
Carbon (C)	20	14	kg
Hydrogen (H)	10	7	kg
Nitrogen (N)	3	2.1	kg
Calcium (Ca)	1.5	1	kg
Phosphorus (P)	1	0.7	kg
Potassium (K)	0.25	170	g
Sulphur (S)	0.2	140	g
Chlorine (Cl)	0.1	70	g
Sodium (Na)	0.1	70	g
Magnesium (Mg)	0.04	30	g
Iron (Fe)	0.004	3	g
Copper (Cu)	0.0005	300	mg
Manganese (Mn)	0.0002	100	mg
Iodine (J)	0.00004	30	mg

Table 4.1.4: Composition of the Body of an Adult Human

For composition in terms of the chemical elements, see Table 4.1.3; a comparison between men and women is found in Table 4.1.5.

Figures after MITCHELL 1945

	Relative body weight (%)	Water (%)	Fat (%)	Protein (%)	Ash (%)
Skin	7.81	64.68	13.00	22.10	0.68
Skeleton	14.84	31.81	17.18	18.93	28.91
Teeth	0.06	5.00	0.00	23.00	70.90
Striated musculature	31.56	79.52	3.35	16.50	0.93
Brain, bone marrow	2.52	73.33	12.68	12.06	1.37
Liver	3.41	71.46	10.35	16.19	0.88
Heart	0.69	73.69	9.26	15.88	0.80
Lungs	4.15	83.74	1.54	13.38	0.95
Spleen	0.19	78.69	1.19	17.81	1.13
Kidneys	0.51	79.47	4.01	14.69	0.96

(Cont.) →

	Relative body weight (%)	Water (%)	Fat (%)	Protein (%)	Ash (%)
Pancreas	0.16	73.08	13.08	12.69	0.93
Intestine	2.07	79.07	6.24	13.19	0.86
Adipose tissue	13.63	50.09	42.44	7.06	0.51
All other tissue	13.63	70.40	12.39	16.06	1.01
Blood, lymph	3.79	93.33	0.17	5.68	0.94
Total		67.85	12.51	14.39	4.84

Table 4.1.5: Relative Share of Individual Tissues in Body Composition of Men and Women

From SCHWIDETZKY 1965

	♀ (%)	♂ (%)
Musculature	35.8	41.8
Fat	28.2	18.2
Skeleton	15.1	15.9
Glands and remaining tissue	20.9	24.1

Table 4.1.6: Growth of the Human Embryo During Pregnancy

Figures from *Documenta Geigy* 1975, LACHMANN 1976, MEYER 1964

Age	Somites (n)	Length (Crown/ rump)	Length (Crown/ heel)	Weight (g)
20 days	1– 4			
21 days	4– 7	2 mm		
22 days	7–10			
23 days	10–13			
24 days	13–17			
25 days	17–20			
26 days	20–23			
27 days	23–26	4 mm		
28 days	26–29			0.02 g
30 days	34–35			

Age	Somites (n)	Length (Crown/ rump)		Length (Crown/ heel)		Weight (g)	
35 days	42–44	5–8	mm				
6 weeks		10–14	mm				
7 weeks		17–22	mm	1.9	cm		
8 weeks		28–30	mm	3.0	cm	1	g
12 weeks		5.6	cm	7.3	cm	14	g
16 weeks		11.2	cm	15.7	cm	105	g
20 weeks		16.0	cm	23.9	cm	310	g
24 weeks		20.3	cm	29.6	cm	640	g
28 weeks		24.2	cm	35.5	cm	1,080	g
32 weeks		27.7	cm	40.9	cm	1,670	g
36 weeks		31.3	cm	45.8	cm	2,400	g
40 weeks (birth)		35.0	cm	50.0	cm	3,300	g

Table 4.1.7: Normal Growth of Boys and Girls

Figures given for height and weight. Numbers represent the range covering 80% of all individuals. The shortest documented adult was Pauline Musters, at a height of 0.48 m; the tallest was R.P. Wadlow, at 2.72 m.

Figures from *Documenta Geigy* 1975, Schmidt 1984

Age	Boys		Girls	
	Height (cm)	Weight (kg)	Height (cm)	Weight (kg)
Birth	49.0– 52.5	2.80– 3.90	47.0– 52.0	2.70– 3.96
1 month	52.5– 56.8	3.57– 4.58	51.4– 55.5	3.44– 4.42
3 months	59.0– 63.2	5.40– 6.66	58.0– 61.6	4.89– 6.23
6 months	65.1– 71.3	7.00– 8.77	64.0– 69.3	6.53– 8.27
9 months	71.0– 75.0	8.30–10.36	68.6– 73.5	7.62– 9.71
1 year	73.4– 79.0	9.71–11.13	72.0– 77.0	8.40–11.05
18 months	79.9– 86.0	10.36–12.91	78.5– 85.0	10.15–12.42
2 years	81.7– 92.3	9.70–15.10	81.4– 91.6	10.10–13.70
3 years	91.0–101.6	11.90–17.30	89.6–100.8	11.70–16.90
4 years	97.9–109.1	14.00–19.40	96.5–108.1	13.30–19.30
5 years	104.2–116.2	16.10–21.70	103.5–114.9	15.20–21.40
6 years	110.2–122.8	18.00–24.00	109.4–122.4	16.80–24.40
7 years	115.8–129.0	19.80–26.60	115.3–129.5	18.40–28.60
8 years	121.3–134.7	20.80–29.80	121.7–134.7	20.70–32.70
9 years	126.7–139.9	21.70–33.30	126.5–139.9	22.60–35.00
10 years	131.2–145.4	23.50–37.50	130.2–144.8	24.80–38.20

(Cont.) →

Age	Boys		Girls	
	Height (cm)	Weight (kg)	Height (cm)	Weight (kg)
11 years	134.9–150.5	27.10–42.90	133.4–150.2	26.40–43.20
12 years	138.2–155.8	30.60–48.40	138.7–158.3	29.40–48.60
13 years	143.3–163.5	34.20–54.00	144.8–163.4	33.60–55.00
14 years	150.2–171.0	38.30–58.90	150.8–165.6	39.40–60.80
15 years	157.1–176.5	42.50–63.70	153.8–168.4	44.30–62.10
16 years	163.3–180.1	47.40–68.00	155.3–170.5	46.40–63.80
17 years	166.7–181.9	52.10–69.50	155.8–171.8	47.70–65.30

Table 4.1.8: Number of Bones in the Human Body

After BERTELSMANN 1979, MEYER 1964, RUCKER 1967

Skull	25
Spine	34
Rib cage	25
Shoulder girdle	4
Upper arm	2
Forearm	4
Carpal (wrist) bones	16
Metacarpal bones	10
Fingers	28
Pelvis	6
Thigh	2
Lower leg	4
Tarsal bones	14
Metatarsal bones	10
Toes	28
Total	212

Table 4.1.9: Dimensions and Weight of Selected Human Organs

Figures represent mean values.

Data after BERTELSMANN 1979, MEYER 1964, RUCKER 1967

	Length		Width		Circumference		Weight	
Heart	15	cm	9	cm	25	cm	300	g
Lung	25	cm					1,100–1,200	g
Stomach (empty)	25–35	cm	12–15	cm			300	g
Liver	30–35	cm	20–25	cm			1,200–1,700	g
Pancreas	12–15	cm	8– 9	cm			70– 80	g
Kidney	10–12	cm	6– 7	cm			120– 160	g
Brain							1,200–1,500	g
Eyeball	2.4	cm	2.4	cm	7.4	cm	7.4	g

Table 4.1.10: Specific Weight of Selected Human Organs

Figures from OPPENHEIMER and PINCUSSEN 1925

Organ	Specific weight
Bone	1.936
Spongiosa	1.825–1.901
Musculature	1.058
Nerves	1.035
Brain and bone marrow	1.03
Liver	1.05
Kidney	1.05
Adipose tissue	0.924–0.932
Human as a whole	1.027–1.059

Table 4.1.11: Water Content of Human Organs and Tissue

After ALTMAN and DITTMER 1973, BERTELSMANN 1979, HEIDERMANNS 1957, MITCHELL et al. 1945

Organ/Tissue	Water content (%)
Brain	77
White matter	70
Gray matter	84
Bone marrow	71
Peripheral nerve	66
Striated muscle	79.5
Skin	64.7

(Cont.) →

Organ/Tissue	Water content (%)
Hair	4.1
Bone	43.9–55
Tooth. overall	9.2
Dental enamel	2.8
Dentin	11.1
Cartilage	75
Elastic tissue	50
Adipose tissue	15
Liver	71.5
Kidney	79.5
Heart	73.7
Spleen	78.7
Lung	83.7
Digestive tract	79.1
Uterus	79.9
Blood	80
Blood cells	65
Lens of the eye	67.6
Sclera	65
Vitreous humor	99
Cornea	75–80

Table 4.1.12: Temperature of Selected Organs and Parts of the Human Body

After BERTELSMANN 1979, MEYER 1964

Lung	35.2–35.6 °C
Oral cavity	36.5 °C
Auditory canal	36.7 °C
Superior vena cava	36.8 °C
Armpit (axilla)	36.9 °C
Stomach	37.0–37.3 °C
Rectum	37.2 °C
Gluteus muscle	37.7 °C
Inferior vena cava	38.1 °C
Pectoralis major	38.3 °C
Left atrium	38.6 °C
Aorta	38.7 °C
Right atrium	38.8 °C
Liver	41.3 °C

Table 4.1.13: pH-Values of Human Bodily Fluids

Figures represent mean values.

Figures from *Dokumenta Geigy* 1975

Cerebrospinal fluid	7.349
Synovial fluid (knee joint)	7.434
Arterial blood (adult)	7.39
Venous blood (adult)	7.398
Lymph	7.4
Saliva	6.4
Gastric juice, men	1.92
Gastric juice, women	2.59
Bile	7.15–7.5
Feces	7.15
Urine	5.7–5.8
Sweat	4–6.8
Sperm	7.19

4.2 Skin, Skin Adnexa (Hair, Nails, Glands)

Table 4.2.1: Human Skin in Numbers

Figures from BERTELSMANN 1979, *Documenta Geigy* 1975, MEYER 1964, RUCKER 1967

Surface area	1.8 m^2
Thickness	up to 9 mm
Weight	5.5 kg
Number of cells	110,000,000,000
	(= 6,000,000 per cm^2)
Number of sebaceous glands	250,000 (= 15 per cm^2)
Number of perspiratory glands	1,800,000 (= 100 per cm^2)
Number of Ruffini's corpuscles (heat)	40,000
Number of Krause's end bulbs (cold)	240,000
Number of Vater-Pacini corpuscles (pressure, vibration)	40,000
Number of Meißner's tactile corpuscles	500,000
Number of Merkel's disks (touch)	60,000,000
Number of free nerve endings	4,000,000

Table 4.2.2: Number of Human Hairs

For the distribution of hair across the body, see Table 4.2.3.

After BERTELSMANN 1979, MEYER 1964, RUCKER 1967

	Hairs, approx. (n)
1. Head hair	
A blonde person has	150,000
A brown-haired person has	110,000
A black-haired person has	100,000
A red-haired person has	90,000
In addition, all have	
2. Body hairs	25,000
3. Eyelashes	420
4. Eyebrows	600

Table 4.2.3: Number of Hairs per cm² on Various Parts of the Body

Values from BERTELSMANN 1979, MEYER 1964, OPPENHEIMER and PINCUSSEN 1925

At the part in the hair	300–320
Back of the head	200–240
Forehead	200–240
Chin	44
Breast	9
Mons pubis	30–35
Forearm	24
Back of the hand	18
Upper arm	16
Thigh	15
Kneecap	22
Lower leg, calf	9

Table 4.2.4: Hair Diameters (Averages) in an Adult Male

Figures from OPPENHEIMER and PINCUSSEN 1925

Head hairs, overall	0.071 mm
Temples	0.066 mm
Crown of the head	0.067 mm
Forehead	0.069 mm
At the part in the hair	0.075 mm
Back of the neck	0.056 mm
Mustache	0.115 mm
Cheek whiskers	0.104 mm
Chin whiskers	0.125 mm
Eyebrows	0.080 mm
Eyelashes	0.076 mm
Tactile hairs	0.056 mm
Underarm hairs	0.077 mm
Pubic hairs	0.121 mm

Table 4.2.5: Speed of Hair and Nail Growth

Data after ALTMAN and DITTMER 1972, BERTELSMANN 1979

Head hair	0.35	mm/day
Facial hair	2.1–3.5	mm/week
Eyebrows	0.16	mm/day
Underarm hair	0.3	mm/day
Arm hair	1.5	mm/week
Thigh hair	0.2	mm/day
Fingernails	0.086	mm/day
Thumbnail	0.095	mm/day
Toenail	0.004	mm/day
Nail of the big toe	0.006	mm/day

Table 4.2.6: Number of Perspiratory Glands

Figures represent the number of glands per square centimeter of skin surface in different parts of the body. Overall, there are approximately 1.8 million perspiratory glands in the body.

From *Documenta Geigy* 1975, MEYER 1964

	Number/cm^2
Back	55
Buttocks	55
Cheeks	75
Leg	80
Top of the foot	125
Forearm (top)	150
Abdomen	155
Breast	155
Forearm (underside)	160
Forehead	170
Throat	185
Back of the hand	200
Sole of the foot	350–400
Palm of the hand	375–425
Elbow	750

4.3 Heart, Circulation, Blood

Table 4.3.1: Numbers on the Heart and Heartbeat of Humans

Figures represent mean values. For time-related figures on heartbeat, see Table 4.3.2.

Figures from BERTELSMANN 1979, BRUGGAIER and KALLUS 1973, MEYER 1964, RUCKER 1967

Length of heart	15 cm
Weight of heart	300 g
Stroke volume	70 cm^3
Minute volume, resting heart	4,900 cm^3
Minute volume, heart under great exertion	up to 30 l
Amount of blood pumped in 70 years	250,000,000 l
Work per beat, left ventricle	≈ 0.8 J
right ventricle	≈ 0.16 J
Daily work of heart	≈ 100,000 J
Power of heart per heartbeat	1/240 kW
Daily power of heart	425 kW
Number of heartbeats in 70 years	3,000,000,000

Table 4.3.2: Temporal Dimensions of the Human Heartbeat

Figures given at a heart frequency of 75 beats/min. As a point of comparison, the comparative data for skeletal muscles is provided. For additional cardiac data, see Table 4.3.1.

Data after GANONG 1972

	Heart (s)	Skeletal muscles (s)
Length of time of a cardiac cycle	0.80	–
Length of time of the systole (or contraction)	0.27	0.04
Length of time of the action potential	0.25	0.005
Absolute refractory period	0.20	0.004
Relative refractory period	0.05	0.003
Length of time of the diastole	0.53	–

Table 4.3.3: Dimensions of the Human Blood Vessel System

From REIN and SCHNEIDER 1971, supplemented

Arteries	Number	Length (cm)	Total length	Diameter (mm)	Total cross-section (cm^2)	Total surface area (cm^2)	Total volume (cm^3)
Aorta	1	40	40 cm	10	0.8	126	30
Major arteries	40	20	8 m	3	3.0	754	60
First-order arterial branches	600	10	60 m	1	5.0	1,884	50
Second-order arterial branches	1,800	1	18 m	0.6	5.0	339	5
Arterioles	40,000,000	0.2	80 km	0.02	125	50,240	25
Capillaries	1,200,000,000	0.1	1,200 km	0.008	600	301,440	60
Venules	80,000,000	0.2	160 km	0.03	570	150,720	110
Second-order venous branches	1,800	1	18 m	1.5	30	848	30
First-order venous branches	600	10	60 m	2.4	27	4,522	270
Major veins	40	20	8 m	6.0	11	1,507	220
Vena cava	1	40	40 cm	12.5	1.2	157	50

Table 4.3.4: Diameter of the Arteries of an Adult Human

Data after OPPENHEIMER and PINCUSSEN 1925

Arteria pulmonalis	24 mm
Aorta ascendens	22.4 mm
Aorta renalis	13.3 mm
Carotis communis dexter	6.7 mm
Subclavia dextra	6.2 mm
Femoralis dextra	6.2 mm
Renalis dextra	5.3 mm
Capillaries	8 µm

Table 4.3.5: Blood Pressure at Different Points in the Human Circulatory System

Figures in millimeters of mercury (mmHg) and, in parentheses, in hectopascals (hPa). For data on blood pressure in animals, see Table 1.6.4.

After REIN and SCHNEIDER 1971

	Systole	Diastole
Aortic root	110 (147)	80 (107)
Aorta descendens	115 (153)	75 (100)
Arteria iliaca	120 (160)	70 (93)
Arteria brachialis	110 (147)	80 (107)
Arterioles	70–30 (93–40)	
Capillaries	30–15 (40–20)	
Veins	15– 0 (20– 0)	

Table 4.3.6: Age and Sex Dependency of Blood Pressure in Humans

Figures are mean values in millimeters of mercury (mmHg) and, in parentheses, in hectopascals (hPa).

From *Documenta Geigy* 1975, modified

Age (years)	Females Systolic	Diastolic	Males Systolic	Diastolic
1	95 (127)	65 (87)	96 (128)	66 (88)
2	92 (123)	60 (80)	99 (132)	64 (85)
3	100 (133)	64 (85)	100 (133)	67 (88)
5	92 (123)	62 (83)	92 (123)	62 (83)

(Cont.) →

| Age | Females | | Males | |
(years)	Systolic	Diastolic	Systolic	Diastolic
10	103 (137)	70 (93)	103 (137)	69 (92)
12	106 (141)	72 (96)	106 (141)	71 (95)
15	112 (149)	76 (101)	112 (149)	75 (100)
20–24	116 (155)	72 (96)	123 (164)	76 (101)
25–29	117 (156)	74 (99)	125 (167)	78 (104)
30–34	120 (160)	75 (100)	126 (168)	79 (105)
35–39	124 (165)	78 (104)	127 (169)	80 (107)
40–44	127 (169)	80 (107)	129 (172)	81 (108)
45–49	131 (175)	82 (109)	130 (173)	82 (109)
50–54	137 (183)	84 (112)	135 (180)	83 (111)
55–59	139 (185)	84 (112)	138 (184)	84 (112)
60–64	144 (192)	85 (113)	142 (189)	85 (113)
65–69	154 (205)	85 (113)	143 (191)	83 (111)
70–74	159 (212)	85 (113)	145 (193)	82 (109)
75–79	158 (211)	84 (112)	146 (195)	81 (108)
80–84	157 (210)	83 (111)	145 (193)	82 (109)
85–89	154 (205)	82 (109)	145 (193)	79 (105)
90–94	150 (200)	79 (105)	145 (193)	78 (104)

Table 4.3.7: Blood Flow Through Selected Human Organs

Values represent cubic centimeters of blood per minute and per 100 g organ weight. See also Table 4.3.8.

Values from HEIDERMANNS 1957, KOLB 1974

Brain	138
Skeletal muscles (at rest)	13
Skeletal muscles (working)	85
Thyroid gland	591
Kidney	150
Adrenal gland	700
Liver	26
Intestine	31

Table 4.3.8: Proportional Blood Flow Through Selected Human Organs

Figures represent the proportion (%) of the heart minute volume flowing through the given organ. See also Table 4.3.7.

Figures from BAUER et al. 1977

Organ	(%)
Lung	100
Heart	100
Heart arteries	5
Brain	15
Musculature	15
Liver and intestine	35
Kidneys	20
Skin and skeleton	10

Table 4.3.9: Composition of Human Blood

From *Enzyklopädie* 1979

	Quantity per 100 cm³ blood		Quantity per 100 g erythrocytes	
Water	81	g	63	g
Hemoglobin	15	g	33	g
Total protein	19	g	35	g
Total fat	560	mg	600	mg
Neutral fats	135	mg	95	mg
Phosphatide	245	mg	350	mg
Total cholesterol	175	mg	–	
Glycogen	5	mg	–	
Gluose	90	mg	75	mg
Lactic acid	15	mg	–	
Nonprotein nitrogen	30	mg	–	
Urea	15	mg	–	
Creatine	3.9	mg	8	mg
Creatinine	0.9	mg	1.8	mg
RNA	64	mg	–	
Sodium	190	mg	42	mg
Potassium	190	mg	370	mg
Calcium	7	mg	2	mg
Magnesium	3.8	mg	6.2	mg
Iron	48	mg	100	mg
Chloride	290	mg	270	mg

(Cont.) →

	Quantity per 100 cm³ blood	Quantity per 100 g erythrocytes
Anorganic phosphate	2.5 mg	4 mg
Total phosphate	35 mg	66 mg
Bicarbonate	220 mg	–

Table 4.3.10: Most Important Proteins in Human Blood Plasma

Molecular weight and concentration in plasma.

After GANONG 1972, GAUER et al. 1971, *Documenta Geigy* 1975, KARLSON 1974

	Molecular weight	Concentration (mg/100 cm³)	Relative concentration (as % of proteins)
Prealbumin + Albumin			55.3
Prealbumin	61,000	10– 40	
Albumin	69,000	3,900–5,500	
α_1 globulins			4.8
Acid α_1 glycoprotein	44,000	70– 110	
α_1 lipoprotein	200,000	290– 770	
α_1 antitrypsin	54,000	200– 400	
α_1 antichymotrypsin	68,000	30– 60	
α_2 globulins			8.2
Ceruloplasmin	132,000	15– 60	
α_2 macroglobulin	720,000	150– 350	
Haptoglobin	100,000	100– 200	
β globulins			11.2
β lipoprotein	2,400,000	220– 740	
Transferrin	80,000	200– 400	
β_{1C} globulin		80– 140	
Fibrinogen	341,000	200– 450	
γ globulins			20.5
γG globulin	150,000	800–1,800	
γA globulin	160,000	90– 450	
γM globulin	900,000	60– 250	
γD globulin		0.3– 40	

Table 4.3.11: Quantity of Hemoglobin (Hb) and Number of Blood Cells in Humans

More precise figures for white blood cells are found in Table 4.3.12.

After MERCK 1966

	g Hb/100 cm^3 blood	Red blood cells (10^6/mm^3)	White blood cells (1,000/mm^3)	Blood platelets (1,000/mm^3)
Males	14–18	4.5–5.5	5–9	250
Females	12–16	4.0–5.0	5–9	250
Newborns	16–26	4.0–6.0	16–22	230
Infants	10–15	4.0–6.0	9–15	280
Children	10–15.5	4.5–5.5	8–12	340

Table 4.3.12: Number of Leukocytes in Humans

Normal values per cubic millimeter. The average number of leukocytes is currently constantly rising among Europeans and Americans. The cause of this is not known.

From MERCK 1966, modified

	Adults		Children		Infants	
Leukocytes per mm^3	5,000–9,000		8,000–12,000		9,000–15,000	
	Absolute (n)	(%)	Absolute (n)	(%)	Absolute (n)	(%)
Rod neutrophils	150–450	3–5	0–1,200	0–10	0–1,500	0–10
Segmented neutrophils	2,500–6,300	50–70	2,000–7,800	23–65	2,250–9,700	25–65
Eosinophils	100–360	2–4	50–600	1–5	90–1,000	1–7
Basophils	0–50	0–1	0–140	0–1	0–300	0–2
Monocytes	100–600	2–6	–1,000	1–6	–3,000	7–25
Lymphocytes	1,000–3,600	20–40	2,000–6,000	25–50	1,500–10,000	20–70

Table 4.3.13: Composition of Lymph and Blood Plasma in Humans – A Comparison

From FLOREY 1970

	Blood plasma	Lymph	
Protein	6.85	2.61	%
Sugar	123	124	mg/100 cm^3
Nonprotein nitrogen	27.2	27	mg/100 cm^3
Urea	22	23.5	mg/100 cm^3
Amino acid nitrogen	4.9	4.8	mg/100 cm^3
Calcium	10.4	9.2	mg/100 cm^3
Chloride	392	413	mg/100 cm^3

Table 4.3.14: Percentual Distribution of Blood Groups in Selected Peoples

Data after BERTELSMANN 1979, MEYER 1964, SCHWIDETZKY 1962

	0	A	B	AB
Germans	39.1	43.5	12.5	4.9
Hungarians	35.7	43.3	15.7	5.3
Finns	34.1	41.0	18.0	6.9
Sinti and Roma	28.5	26.6	35.3	9.6
Persians	37.9	33.3	22.2	6.6
Eskimos	54.2	38.5	4.8	2.0
Sioux (North Am. Indians)	91.0	7.0	2.0	0.0
Navajo (North Am. Indians)	72.6	26.9	0.2	0.5
Cayapa (South Am. Indians)	100.0	0.0	0.0	0.0
Australians	53.2	44.7	2.1	0.0
Bushmen	56.0	33.9	8.5	1.6
Bantu	63.2	16.4	17.4	3.0
Ainu	17.0	31.8	32.4	18.4
Papuans	62.7	18.5	16.7	2.1

4.4 Respiration

Table 4.4.1: Numbers on the Human Lung and Respiration

Data after MEYER 1964, RUCKER 1967

Length of a lung	25	cm
Weight of a lung	1,200	g
Number of pulmonary alveoli	850,000,000	
Interior surface area	90	m^2
Lung capacity		
Vital capacity	4.1	l
Inspiratory capacity	0.5	l
Expiratory reserve volume	1.8	l
Inspiratory reserve volume	1.8	l
Daily oxygen intake	700	l
Length of capillaries in the alveoli	13,000	m

Table 4.4.2: Age and Sex Dependency of Vital Capacity in Humans

Vital capacity is defined as the total amount of air that is set in motion by one breath, given maximum inhalation and exhalation. Figures represent mean values.

After *Documenta Geigy* 1975

Age (years)	Females (cm^3)	Males (cm^3)
4	717	855
5	959	1,001
8	1,513	1,585
10	1,806	2,022
12	2,217	2,357
14–19	3,330	4,030
20–29	3,600	5,440
30–39	3,460	5,030
40–49	3,680	4,530
50–59	3,250	4,650
60–64	3,140	3,860

Table 4.4.3: Average Air and Energy Consumption of an Adult During Selected Activities

Figures represent consumption per hour.

Data after AURICH 1978

Activity	Air required (liters)	Energy required (kcal) / (kJ)
Sleeping	280	65 / 273
Lying down	400	77 / 323
Standing	450	
Walking	1,000	200 / 840
Biking	1,400	
Swimming	2,600	500 / 2,100
Mountain climbing	3,100	1,000 / 4,200
Rowing	3,600	

Table 4.4.4: Changes in O_2 and CO_2 During Human Respiration

Figures from KEIDEL 1973, REIN and SCHNEIDER 1971

	O_2 (%)	Oxygen partial pressure (pO_2) (mmHg / hPa)	CO_2 (%)	Carbon dioxide partial pressure (pCO_2) (mmHg) / (hPa)
Inspired air	20.9	150 / 200	0.03	0.2 / 0.3)
Alveolar air	14.0	100 / 133	5.6	40 / 53
Expired air	16.0	116 / 155	4	29 / 39
Arterial blood		95 / 127		40 / 53
Venous blood		40 / 53		46 / 61

Table 4.4.5: Age and Sex Dependency of Human Basal Metabolic Rate

Figures are given in kilocalories (and kilojoules) per square meter body surface area per hour.

After *Documenta Geigy* 1975

Age (years)	Males (kcal/m^2 per h) / (kJ/m^2 per h)	Females (kcal/m^2 per h) / (kJ/m^2 per h)
1	53.0 / 222	53.0 / 222
2	52.4 / 219	52.4 / 219
5	49.3 / 206	48.4 / 203
8	46.3 / 194	43.8 / 183

Age (years)	Males (kcal/m² per h) / (kJ/m² per h)	Females (kcal/m² per h) / (kJ/m² per h)
10	44.0 / 184	42.5 / 178
12	42.5 / 178	41.3 / 173
15	41.8 / 175	37.9 / 159
20	38.6 / 162	35.3 / 148
25	37.5 / 157	35.2 / 147
30	36.8 / 154	35.1 / 147
40	36.2 / 152	34.9 / 146
50	35.8 / 150	33.9 / 142
60	34.9 / 146	32.7 / 137
70	33.8 / 141	31.7 / 133
75 and older	33.2 / 139	31.2 / 131

Table 4.4.6: Oxygen Consumption and Heat Generation in Human Organs

Figures refer to a person at rest. On the blood flow through organs, see Tables 4.3.7 and 4.3.8.

Data after GAUER et al. 1971, HALTENORTH 1977, REIN and SCHNEIDER 1971

Organ	Oxygen intake (cm³) Organ per min	100 g organ per min	Organ weight as % of body-weight	Oxygen consump-tion of organ as % of total consump-tion	Heat generation (cal) Organ per min	100 g organ per min
Skin	5.0	0.10	7.0	1	24	0.48
Muscle	43.0	0.16	41.0	18	203	0.75
Brain	43.0	3.20	2.0	19	208	15.40
Lung	12.0	2.00	0.8	5	57	9.50
Heart	29.0	10.00	0.5	14	138	48.00
Kidneys	20.3	7.00	0.4	6	100	34.50
Stomach-Intestine (+ liver, spleen, pancreas)	90.0	3.60	4.0	22	435	17.40

4.5 Nutrition, Digestion, Excretion

Table 4.5.1: Standard Food Requirements for Adults

Figures for the Federal Republic of Germany (FRG) and recommendations of the FAO (Food and Agriculture Organization of the United Nations) for persons 25 years of age. See also Table 4.5.2.

After ALTMAN and DITTMER 1974

						Vitamins			Minerals	
	Weight (kg)	Calo-ries	Protein (g/day)	Thia-mine (mg/ day)	Ribo-flavin (mg/ mg/ day)	Niacin (mg/ day)	Vit. C (mg/ day)	Vit. A (IE/ day)	Ca (g/day)	Fe (mg/ day)
FRG										
♂	72	2,550	72	1.7	1.8	18	75	5,000	0.8	10
♀	60	2,200	60	1.5	1.8	14	75	5,000	0.8	12
FAO										
♂	65	3,200	46	1.3	1.8	21.1	–	–	0.4–0.5	–
♀	55	2,300	39	0.9	1.3	15.2	–	–	0.4–0.5	–

Table 4.5.2: Recommended Daily Protein and Overall Energy Intake in the Human Diet

Figures for protein intake represent grams of protein per kilogram of body weight and per day (± 20 %); figures for overall energy intake are given in kilocalories and kilojoules, calculated for a primarily sedentary activity. For breast-feeding women, 5 g protein should be added to the figures given for every 100 g of breast-fed milk. See also Table 4.5.1.

After *Verbraucherdienst* 1979

	Protein intake (g) ♂	♀	Energy intake (kcal) / (kJ) ♂	♀
Adults (age)				
25 years	0.9	0.9	2,600 / 10,900	2,200 / 9,200
45 years	0.9	0.9	2,400 / 10,000	2,000 / 8,400
65 years	1.0	1.0	2,200 / 9,200	1,800 / 7,500

	Protein intake (g) ♂	♀	Energy intake (kcal) / (kJ) ♂	♀
Pregnant women				
5 months or less		0.9		2,200 / 9,200
More than 5 months		1.5		2,600 / 10,900
Breast-feeding women		0.9		2,800 / 11,700
Children				
6 months	2.5		600 / 2,500	
6–12 months	2.2		900 / 3,800	
1 to < 4 years	2.2		1,200 / 5,000	
4 to < 7 years	2.0		1,600 / 6,700	
7 to < 10 years	1.8		2,000 / 8,400	
10 to < 13 years	1.5	1.4	2,400 / 10,000	2,100 / 8,800
13 to < 15 years	1.5	1.4	2,700 / 11,300	2,400 / 10,000
15 to < 19 years	1.2	1.0	3,100 / 13,000	2,500 / 10,500

Table 4.5.3: Human Digestive Capacity for Selected Nutrients

Figures represent the digestive capacities of a normal person (70 kg body weight) whose basal metabolic rate is ca. 1,700 kcal (7,200 kJ) daily and whose energy requirements are ca. 2,400 kcal (10,000 kJ) daily.

Data after FÖRSTER 1980, modified

Nutrient	Digestive capacity	Energy content (kcal) / (kJ)
Carbohydrates		
Starch	1,000 g	3,800 / 16,000
Glucose oligosaccharide	2,000 g	7,600 / 32,000
Saccharose	800 g	3,200 / 13,500
Lactose	200 g	760 / 3,200
Fats		
Long-chain triglycerides	500 g	4,050 (17,000)
Proteins		
High-grade protein	400 g	1,700 (7,000)
Oligopeptides	180– 250 g	710– 950 (3,000–4,000)

Table 4.5.4: Components of a Chemically Defined Diet (CDD) that Meets Human Nutritional Needs

The amounts given here are the equivalent of 300 kcal (= 1,255 kJ). On human vitamin requirements, see also Table 4.5.7. IU = international units

From GRIMME 1975, modified

Amino acids		
L-Isoleucine	0.275	g
L-Leucine	0.435	g
L-Lysine monohydrochloride	0.327	g
L-Methionine	0.282	g
L-Phenylalanine	0.313	g
L-Threonine	0.275	g
L-Tryptophan	0.085	g
L-Valine	0.303	g
L-Alanine	0.293	g
L-Arginine hydrochloride	0.537	g
L-Aspartic acid	0.625	g
L-Glutamine	1.032	g
Glycine	0.478	g
L-Histidine monohydrochloride monohydrate	0.133	g
L-Proline	0.329	g
L-Serine	0.202	g
L-Tyrosine ethylester hydrochloride	0.343	g
	6.330	g

Carbohydrates		
Glucose monohydrate	24.000	g
Oligosaccharide (from enzymatically and hydrolytically decomposed cornstarch)	45.500	g
	69.500	g

Essential lipid acids		
Safflower oil	0.222	g

Vitamins and other substances			
Vitamin A acetate	0.287	mg	(834 IU)
Vitamin C	11.700	mg	
Vitamin D_2	1.670	mg	(67 IU)
Vitamin E acetate	3.330	mg	
Vitamin B_1 (aneurine chloride hydrochloride)	0.200	mg	
D-Biotin	0.033	mg	
Calcium D-pantothenate	1.670	mg	
Choline hydrogen tartrate	14.170	mg	
Vitamin B_{12} (cyanocobalamin)	0.830	µg	
Folic acid	0.017	mg	
Inositol	19.420	mg	

Nicotinamide	2.220	mg
Vitamin B_6 (pyridoxine hydrochloride)	0.330	mg
Vitamin B_2 (riboflavin 5-phosphate ester, sodium salt)	0.200	mg
Vitamin K_1 (phytomenadione)	0.020	mg
Sodium glycerophosphate	430.000	mg
Glucono-δ-lactone	1.060	g
Sorbic acid	150.000	mg

Minerals and trace elements

Sodium chloride	978.000	mg
Potassium chloride	672.000	mg
Calcium chloride	489.000	mg
Magnesium oxide	43.200	mg
Ferrous sulfate [iron (II) sulfate]	8.260	mg
Copper (II) acetate	1.030	mg
Manganese (II) acetate	2.090	mg
Zinc chloride	0.140	mg
Potassium iodide	0.030	mg

Table 4.5.5: Energy Requirements in Humans for Selected Physical Activities

Figures represent the amount exceeding the basal metabolic rate required per hour.

After FÖRSTER 1980

Activity	Energy requirement per hour (kcal)	(kJ)
Walking	50– 200	200– 800
Biking	100– 300	400–1,200
"Fitness path"*	200– 400	800–1,600
Endurance run	200– 400	800–1,600
Dancing	300– 400	1,200–1,600
Soccer	400– 500	1,600–2,000
Tennis	400– 500	1,600–2,000
Swimming	200– 600	800–2,400
Jogging	500–1,000	2,000–4,000
Skiing	400– 800	1,600–3,200

*Forest paths set up with calisthenics stations

Table 4.5.6: Composition and Energy Content of Selected Foodstuffs

Calculated for 100 g edible substance.
For the energy values of plant parts, see Table 2.5.8.

From *Documenta Geigy* 1975, modified

	Water	Proteins	Fats	Carbo-hydrates	Energy content (kcal)	(kJ)
Apples, sweet	84.0	0.3	0.6	15.0	58	243
Bananas	75.7	1.1	0.2	22.2	85	356
Strawberries	89.9	0.7	0.5	8.4	37	155
Cherries	83.4	1.2	0.4	14.6	60	251
Oranges	87.1	1.0	0.2	12.2	49	205
Plums	85.7	0.7	0.1	12.3	50	210
Raisins	18.0	2.5	0.2	77.4	289	1,210
Grapes	81.4	0.6	0.3	17.3	67	280
Watermelon	92.6	0.5	0.2	6.4	26	109
Rhubarb	94.9	0.5	0.1	3.8	16	67
White kidney bean	11.6	21.3	1.6	61.6	338	1,415
Cucumbers	95.6	0.8	0.1	3.0	13	55
Carrots	88.6	1.1	0.2	9.1	40	167
Potatoes	79.8	2.1	0.1	17.7	76	318
White cabbage	92.1	1.4	0.2	5.7	25	105
Cauliflower	91.0	2.7	0.2	5.2	27	113
Brussel sprouts	84.8	4.7	0.4	8.7	47	197
Sauerkraut	92.8	1.0	0.2	4.0	18	75
Leek	87.8	2.0	0.3	9.4	44	184
Pepper (green)	92.8	1.2	0.2	5.3	24	100
Radish	93.7	1.1	0.1	3.6	18	75
Red beets	87.3	1.6	0.1	9.9	43	180
Lettuce	95.1	1.3	0.2	2.5	14	59
Celeriac	88.4	1.8	0.3	8.5	40	167
Asparagus	92.9	2.1	0.2	4.1	21	88
Spinach	90.7	3.2	0.3	4.3	26	109
Tomatoes	93.5	1.1	0.2	4.7	22	92
Onions	89.1	1.5	0.1	8.7	38	159
Agaricus mushrooms	90.8	2.8	0.2	3.7	22	92
Porcini mushrooms	88.6	2.8	0.4	5.9	31	130
Baker's yeast	71.0	12.1	0.4	11.0	87	360
Peanuts (roasted)	1.8	26.2	48.7	20.6	582	2,436
Hazelnuts	6.0	12.7	60.9	18.0	627	2,624
Coconuts	48.0	4.2	34.0	12.8	351	1,469
Walnuts	3.5	14.8	64.0	15.8	651	2,725
Almonds	4.7	18.6	54.2	19.5	598	2,503

	Water	Proteins	Fats	Carbo-hydrates	Energy content (kcal)	(kJ)
Wheat flour	12.6	12.1	2.1	71.5	331	1,385
Rye flour	14.3	10.8	1.5	71.8	310	1,297
Wheat germ	11.5	26.6	10.9	46.7	363	1,519
Oatmeal	10.3	13.8	6.6	67.6	387	1,620
Bread roll	34.0	6.8	0.5	58.0	269	1,126
Rye bread	38.5	6.4	1.0	52.7	227	950
Spaghetti	10.4	12.5	1.2	75.2	369	1,544
Honey	17.2	0.3	0.0	82.3	304	1,272
Marzipan	8.8	8.0	18.0	64.0	428	1,792
Milk chocolate	0.9	7.7	32.3	56.9	520	2,176
Dextrose	0.0	0.0	0.0	99.5	385	1,611
Butter	17.4	0.6	81.0	0.7	716	2,996
Cod-liver oil	0.0	0.0	99.9	0.0	901	3,771
Margarine	19.7	0.5	78.4	0.4	698	2,921
Lard	1.0	0.0	99.0	0.0	901	3,771
Chicken egg	74.0	12.8	11.5	0.7	162	678
Cow milk	88.5	3.2	3.7	4.6	64	268
Whipped cream	64.1	2.2	30.4	2.9	288	1,205
Yogurt	86.1	4.8	3.8	4.5	71	297
Low-fat quark (curd cheese)	79.4	17.2	0.6	1.8	86	360
Camembert	51.3	18.7	22.8	1.8	287	1,201
Emmentaler	34.9	27.4	30.5	3.4	398	1,666
Duck	54.0	16.0	28.6	0.0	326	1,365
Roast chicken	72.7	20.6	5.6	0.0	138	578
Rabbit	73.0	22.3	0.9	0.2	103	431
Calf's liver	70.7	19.2	4.7	4.1	140	586
Veal chop	70.0	19.5	9.0	0.0	164	686
Lamb chop	52.0	14.9	32.0	0.0	352	1,473
Fillet of beef	75.1	19.2	4.4	0.0	122	511
Pork	71.2	18.6	9.9	0.0	168	703
Pork chop	53.9	15.2	30.6	0.0	341	1,427
Cervelat sausage	55.6	12.5	27.6	1.8	256	1,072
Veal sausage	65.2	11.1	21.7	0.0	241	1,009
Trout	77.6	19.2	2.1	0.0	101	423
Herring	62.8	17.3	18.8	0.0	243	1,017
Carp	72.4	18.9	7.1	0.0	145	607
Haddock	80.5	18.3	0.1	0.0	79	331
Tuna	52.5	23.8	20.9	0.0	290	1,214

(Cont.) →

	Water	Proteins	Fats	Carbo-hydrates	Energy content (kcal)	(kJ)
Oysters	83.0	9.0	1.2	4.8	68	285
Mussels	84.1	11.7	1.9	2.2	76	318
Edible snails	82.0	15.0	0.8	2.0	75	314
Cuttlefish	82.2	15.3	0.8	0.0	73	306
Shrimps	78.2	18.7	2.2	0.0	97	406
Lobster	78.5	16.9	1.9	0.5	91	381

Table 4.5.7: Daily Human Vitamin Requirements

For the vitamin content of foodstuffs, see Table 4.5.8.

From KARLSON 1980, modified

A	Retinol	1.5– 2	mg
D	Calciferol	0.02	mg
E	Tocopherol	20	mg
K	Menaquinone	1	mg[a,b]
B_1	Thiamine	1.4	mg
B_2	Riboflavin	1.7	mg
B_3	Nicotinamide	20	mg[a,b]
B_5	Pantothenic acid	10	mg[a]
B_6	Pyridoxine	2	mg[a]
B_9	Folic acid	0.4	mg[a]
B_{12}	Cobalamin	0.005	mg
C	Ascorbic acid	60	mg
H	Biotin	0.25	mg[a]

[a]Estimated value, [b]synthesized by intestinal bacteria

Table 4.5.8: Vitamin Content of Selected Foodstuffs

Figures refer to 100 g edible substance. For daily human vitamin requirements, see Table 4.5.7. IU = international units, t = only present as trace element. (B_1 = Thiamine, B_2 = Riboflavin, B_3 = Nicotinamide, B_5 = Panthothenic acid, B_6 = Pyridoxine, B_9 = Folic acid, B_{12} = Cabolamin

Figures from *Documenta Geigy* 1975

	A (IU)	B_1 (mg)	B_2 (mg)	B_3 (mg)	B_5 (mg)	B_6 (mg)	B_9 (mg)	B_{12} (mg)	C (mg)	D (IU)	E (mg)	Biotin (mg)	K (mg)
Apples, sweet	90	0.04	0.02	0.1	0.1	0.03	0.002	–	5	–	0.3	0.001	–
Bananas	190	0.05	0.06	0.6	0.2	0.32	0.01	–	10	–	0.2	0.004	–
Blackberries	200	0.03	0.04	0.4	0.25	0.05	0.012	–	21	–	–	0.0004	–
Strawberries	60	0.03	0.07	0.6	0.26	0.04	0.005	–	60	–	–	0.0011	–
Grapefruit juice	10	0.03	0.02	0.2	0.16	0.014	0.001	–	45	–	–	0.0007	–
Elderberries	600	0.07	0.08	1.5	0.18	0.25	0.017	–	18	–	–	0.002	–
Red currants	120	0.04	0.02	0.3	0.06	0.05	–	–	41	–	–	0.0026	–
Black currants	220	0.05	0.03	0.3	–	0.08	–	–	136	–	–	0.0026	–
Cherries	1,000	0.05	0.06	0.3	0.08	0.05	0.006	–	10	–	–	0.0004	–
Oranges	200	0.1	0.03	0.2	0.2	0.03	0.005	–	50	–	0.23	0.01	–
Lemons	20	0.04	0.02	0.1	0.2	0.06	0.007	–	45	–	–	–	–
Carrots	11,000	0.06	0.06	0.6	0.27	0.12	0.008	–	2–10	–	0.45	0.003	0.08
Potatoes	5	0.11	0.04	1.2	0.3	0.2	0.006	–	20	–	0.06	0.0001	–
White cabbage	70	0.05	0.04	0.32	0.26	0.11	0.08	–	46	–	0.7	–	–
Cress	9,300	0.08	0.26	1.0	–	–	–	–	69	–	–	–	–
Parsley	8,500	0.12	0.26	1.2	0.03	0.2	0.04	–	172	–	–	0.0004	–
Lettuce	970	0.06	0.07	0.3	0.1	0.07	0.02	–	8	–	0.6	0.003	–
Asparagus	900	0.18	0.20	1.5	0.62	0.14	0.11	–	33	–	2.5	–	–
Spinach	8,100	0.1	0.2	0.6	0.3	0.2	0.075	–	51	–	2.5	0.007	0.04
Tomatoes	900	0.06	0.04	0.6	0.31	0.1	0.008	–	23	–	0.27	0.007	–
Onions	40	0.03	0.04	0.2	0.17	0.1	0.01	–	10	–	0.26	0.004	–

(Cont.) →

	A (IU)	B_1 (mg)	B_2 (mg)	B_3 (mg)	B_5 (mg)	B_6 (mg)	B_9 (mg)	B_{12} (mg)	C (mg)	D (IU)	E (mg)	Biotin (mg)	K (mg)
Agaricus mushrooms	–	0.1	0.44	6.2	2.1	0.05	0.03	–	5	150	0.83	0.016	–
Walnuts	30	0.3	0.13	1.0	0.7	1.0	0.077	–	2	–	1.5	0.037	–
Oatmeal	–	0.55	0.14	1.1	0.92	0.75	–	–	–	–	0.25	–	–
Wheat flour	–	0.06	0.05	0.9	–	–	–	–	–	–	–	–	–
Honey	–	t	0.04	0.3	0.06	0.01	0.003	–	1	–	–	–	–
Milk chocolate	270	0.01	0.34	0.3	–	–	–	–	–	–	1.1	–	–
Butter	3,300	t	0.01	0.1	t	t	–	–	t	40	2.4	–	–
Margarine	3,000	–	–	–	–	–	–	–	–	–	–	–	–
Cod-liver oil	85,000	–	–	–	–	–	–	–	–	8,500	26.0	–	0.002
Chicken egg, raw	1,180	0.12	0.34	0.1	1.6	0.25	0.005	0.002	–	200	1.0	0.02	–
Cow milk, fresh	140	0.04	0.15	0.07	0.33	0.05	0.0001	0.0006	1	0.5–4	0.06	0.002	–
Buttermilk	35	0.04	0.18	0.04	0.36	0.04	0.0003	0.0003	1	–	0.05	0.002	–
Camembert	1,010	0.05	0.45	0.25	0.1–0.9	0.25	–	–	0.5	–	–	0.005	–
Emmentaler	1,140	0.05	0.33	0.1	–	0.09	–	–	1	100	0.35	–	–
Low-fat quark (curd cheese)	35	0.04	0.31	0.1	0.1	0.01	0.03	–	–	40	–	0.002	0.15
Calf's liver	22,500	0.28	2.72	17	9.7	1.2	0.05	0.06	32	50	0.9–1.6	0.075	–
Fillet of beef	–	0.1	0.13	4.6	–	–	–	–	–	–	0.5	–	–
Pork chops	–	0.8	0.19	4.3	0.4	0.48	0.002	0.001	–	–	0.6	0.005	–
Pork heart	30	0.43	1.24	6.6	2.5	0.43	–	0.003	3	–	1.4	0.02	–
Pork liver	10,900	0.43	2.7	16.4	7.0	0.85	0.22	0.01	27	–	1.0	0.1	–
Salami	–	0.18	0.2	2.6	–	–	–	–	–	–	0.11	–	–
Mussels	180	0.16	0.22	1.6	–	–	–	0.015	t	–	–	–	–
Oysters	310	0.18	0.23	2.5	0.5	0.11	0.004	–	5	–	–	–	–
Trout	150	0.09	0.25	3.5	–	–	–	–	–	0.0002	0.001	–	–
Cod	0.06	0.07	0.2	0.12	0.001	0.0005	2	2.2	–	–	–	0.0005	–
Tuna (canned)	90	0.05	0.06	10.8	0.2	0.25	0.003	0.001	–	–	–	–	–

Table 4.5.9: Adult Daily Requirements of Essential Amino Acids
See also Table 4.5.4.

After *Documenta Geigy* 1975, Karlson 1974

| Amino acid | Absolute amount (g) | | mmol/kg |
	Males	Females	body weight
Valine	0.80	0.65	0.28
Leucine	1.10	0.62	0.39
Isoleucine	0.70	0.45	0.21
Lysine	0.80	0.50	0.24
Phenylalanine	0.30	0.22	0.18
Tryptophan	0.25	0.16	0.04
Methionine	0.20	0.35	0.07
Threonine	0.50	0.31	0.20

Table 4.5.10: Fatty Acid and Cholesterol Content of Selected Foodstuffs

Unsaturated fatty acids cannot be synthesized by the body and thus must be ingested. The most important of these essential fatty acids is linoleic acid.
Unless otherwise noted, figures are per 100 g.

Data after *Verbraucherdienst* 1979

Foodstuff	Fatty acids, polysaturated (g)	Fatty acids, polyunsaturated (g)	Cholesterol (mg)
Beef, lean	5	2	100
Pork, lean	5	1	90
Venison	1	1	110
Roast chicken	3	1	53
Duck	12	8	75
Brains	4	1	3,146
Kidney	4	1	350
Liver	4	3	250
Ham	20	4	400
Liverwurst	15	5	85
Trout	1	1	70
Herring	3	3	80
Lard	41	10	100
Butter	52	2	230

(Cont.) →

Foodstuff	Fatty acids, polysaturated (g)	Fatty acids, polyunsaturated (g)	Cholesterol (mg)
Margarine	24	24	–
Mayonnaise, 80% fat	12	48	55
1 egg (ca. 60 g)	2	1	280
Egg yolk (from ca. 5 eggs)	10	2	1,400
Milk, 3.5% fat	2	–	11
Cream, 30% fat	21	1	94
Quark (curd cheese), 20% fat i.d.m.	3	–	14
Edamer, 30% fat i.d.m.	9	–	42
Butter cheese, 60% fat i.d.m.	21	1	92
Whole-grain bread, white bread	1	–	–
Pasta	1	1	118
Milk chocolate	20	1	74

Table 4.5.11: Numbers on the Human Digestive System

After MEYER 1964, RUCKER 1967

Number of oral salivary glands	6,000,000
Saliva produced daily	1–2 l
Stomach capacity	3 l
Gastric juice produced daily	1-2 l
Bile produced daily	0.75 l
Pancreatic juice produced daily	1.2 l
Length of the small intestine	3 m
Intestinal juice produced daily	3 l
Number of villi in the small intestine	7,000,000
Total surface area of the villi	40 m^2
Number of microvilli per villus	300,000
Contact surface area that microvilli provide food	2,000 m^2
Number of mucous membrane cells per villus	30,000
Number of mucous membrane cells im small intestine	200,000,000,000

Table 4.5.12: Teeth Eruption Times in Humans

Data after PFLUGFELDER 1962, RUCKER 1967, STARCK 1975

Deciduous (baby) teeth:

Central incisor	Months 6-8
Lateral incisor	Months 7-9
Front molar	Months 12-15
Canine	Months 16-20
Back molar	Months 20-24

Permanent teeth:

Molar	Year 7
Central incisor	Year 8
Lateral incisor	Year 9
Premolar	Year 10
Canine	Years 11-13
Premolar	Years 11-15
2nd molar	Years 13-16
3rd molar (wisdom tooth)	Years 18-30

Table 4.5.13: Numbers on the Human Kidney

After BERTELSMANN 1979, MEYER 1964, RUCKER 1967

Weight of a single kidney	120–160 g
Length of a kidney	10–12 cm
Number of renal corpuscles (glomeruli)	2,000,000
Number of loops of Henle	2,000,000
Total length of the loops of Henle	100 km
Inner surface area of the loops of Henle	20 m^2
Daily blood flow through kidney	1,500 l
Glomerular ultrafiltrate produced daily	180 l
Urine produced daily	0.5–4 l
Salt reabsorbed daily	1 kg
Sodium bicarbonate reabsorbed daily	0.5 kg
Daily renal oxygen consumption	30 l

Table 4.5.14: Filtration and Reabsorption of Selected Substances in the Human Kidney

Figures from KEIDEL 1970, REIN and SCHNEIDER 1971

	Mean concentration in blood plasma (%)	Amount in glomerular ultrafiltrate = daily filtration (g)	Daily reabsorption (g)	Daily excretion output (g)	Mean concentration in urine (%)
Water		170,000	168,500	1,500	
Protein	7–8	11	11	0	0
Glucose	0.1	170	169.5	0.5	0.01
Uric acid	0.0005	8.5	7.97	0.53	0.035
Urea	0.025	46	19	27	1.8
Creatinine	0.001	1.7	0	1.7	0.11
Sodium	0.333	566	561	5	0.333
Calcium	0.01	17	16.8	0.2	0.015
Potassium	0.017	28.9	26.2	2.7	0.18
Chloride	0.373	634	628.7	5.3	0.353
HCO_3	0.159	270	269.7	0.3	0.02

Table 4.5.15: Daily Urine Volume in Humans at Different Ages

Figures represent mean values. Amounts can be greatly influenced by liquid intake, perspiration, and similar phenomena.

Data after *Merck* 1966

	Age (years)	Daily urine volume (cm^3)
Children	1– 3	500– 600
	3– 5	600– 700
	5– 8	650–1,000
	8–14	1,000–1,400
Adults	(> 14)	1,000–1,600

4.6 Brain, Nerves, Senses

Table 4.6.1: Numbers on the Human Nervous System

Figures represent mean values. For further data on the nervous system, see Tables 4.6.2 and 4.6.3.

Values after RUCKER 1964

Weight of the brain	1.5 kg
Weight of the cerebrum	1.25 kg
Number of neurons in the brain	14,000,000,000
Number of interconnections between neurons (synapses)	7,000,000,000,000
Thickness of the cerebral cortex	2–5 mm
Number of neurons in the cerebral cortex	12,000,000,000
Number of interconnections in the cerebrum	1,000,000,000,000
Length of nerve fiber in the cerebrum	500,000 km
Length of nerve fiber outside of the brain	480,000 km
Daily loss of neurons	50,000–100,000

Table 4.6.2: Brain Volumes of Hominids

For the brain weights of animals, see Tables 1.10.1 and 1.10.2.

Figures from HEBERER 1968, v. KÖNIGSWALD 1960, PORTMANN 1959, SALLER 1964, STEITZ 1979

	Range of variation (cm^3)	Mean value (cm^3)
Chimpanzee	320– 480	394
Orangutan	295– 575	411
Gorilla	340– 685	498
Proconsul		500
Australopithecus	450– 750	576
Homo erectus erectus (Pithecanthropus)	835– 900	871
Homo erectus pekinensis (Sinanthropus)	915–1,225	1,046
Homo heidelbergensis	775– 900	
Homo sapiens preneanderthalensis	1,070–1,280	1,175
Steinheim man	1,100–1,200	
Homo sapiens neanderthalensis	1,300–1,610	1,438
Cro-Magnon man (Homo sapiens sapiens)	1,250–1,700	1,400
Present humans (Homo sapiens sapiens)	1,010–2,000	1,400

Table 4.6.3: Brain Weights of Several Eminent Humans

Brain weights are only comparable if those compared died at the same age, since cell division ceases in the human brain after the age of 15 and the brain thereafter continuously loses cellular material.

After BERTELSMANN 1979, MEYER 1964, SLIJPER 1967

	Weight (g)	Age (years)
Bismarck	1,807	83
Lord Byron	2,230	36
Cromwell	2,000	59
Cuvier	1,861	62
Dante	1,420	56
Gauß	1,492	78
Helmholtz	1,420	73
Kant	1,650	80
Liebig	1,350	70
Schiller	1,580	46

Table 4.6.4: Sleep Requirements and Proportion of REM Sleep for Humans of Different Ages

REM sleep is defined as that phase of sleep in which rapid eye movements (REM) and vivid dreams take place.

From GANONG 1972

Age		Daily sleep requirement (h)	Proportion of sleep spent in REM sleep (%)
1–15	days	16	50
3– 5	months	14	40
2– 3	years	13	25
5– 9	years	11	19
10–14	years	9	20
19–30	years	8	22
33–45	years	7	19
50–70	years	6	15

Table 4.6.5: Number of Receptors and Afferent Nerve Fibers of Human Senses

Data after KOLB 1974

Sense	Number of receptors	Number of nerve fibers
Eye	2×10^8	2×10^6
Ear	3×10^4	2×10^4
Taste	10^7	2×10^3
Smell	10^7	2×10^3
Pressure	5×10^5	
Pain	3×10^6	$\times 10^6$
Heat	10^4	
Cold	10^5	

Table 4.6.6: Numbers on the Human Eye and Eyesight

Data after FRANCOIS and HOLLWICH 1977, MEYER 1964, RUCKER 1967

Diameter of the eyeball	24 mm
Weight of the eyeball	7.5 g
Volume of the eyeball	6.5 cm^3
Circumference of the eyeball	74.9 mm
Thickness of the sclera (behind rectus muscles)	0.3 mm
Thickness of the sclera (near optic nerve)	1.35 mm
Thickness of cornea	0.52–0.67 mm
Surface area of the cornea	1.3 cm^2
Diameter of the iris	12 mm
Diameter of the pupil	1.2–9 mm
Lens thickness	4 - 4.14 mm
Lens diameter	9 mm
Lens weight	174 mg
Volume of the vitreous humor	4 cm^3
Weight of vitreous humor	4 g
Thickness of retina at retinal equator	0.18 mm
Retinal thickness at yellow spot (macula lutea)	0.1 mm
Diameter of yellow spot (horizontal)	2 mm
Diameter of yellow spot (vertical)	0.88 mm
Number of visual cells in the retina	132,000,000
Number of rods	125,000,000
Number of cones	7,000,000
Number of interneurons	2,000,000
Number of nerve fibers in the optic nerve	1,000,000
Number of ciliary processes in ciliary body	70–80

(Cont.) →

Length of a ciliary process	2 mm
Thickness of a ciliary process	0.5 mm
Refractive power of the cornea	43 diopters
Refractive power of the lens	19-33 diopters
Resolving power of the eye	$^{1}/_{120}$ arc degree

Table 4.6.7: Changes in the Binocular Horizontal Visual Field of Humans with Age

Figures from ALTMAN and DITTMER 1973

Age (years)	Visual field (degrees) ♂	♀
16–19	174.5	176.4
20–29	174.8	176.2
30–39	173.8	175.3
40–49	171.7	172.7
50–59	167.1	170.1
60–69	160.1	162.3
70–79	151.2	155.5
80 and older	139.8	138.5

Table 4.6.8: Changes in the Near Point of the Human Eye with Age

The near point describes the closest point in front of the eyes at which an object may still be seen clearly given maximum accommodation.

After REIN and SCHNEIDER 1971, SIEBECK 1977

Age (years)	Distance of the near point
10	7 cm
20	9 cm
30	12 cm
40	22 cm
50	40 cm
60	100 cm
70	up to 400 cm
75	∞

Table 4.6.9: Numbers on the Human Ear

See also Table 4.6.10 and, on hearing in animals, see Table 1.11.13.

Figures from BERTELSMANN 1979, MEYER 1964, RUCKER 1976

Diameter of the eardrum	10 mm
Thickness of the eardrum	0.1 mm
Diameter of the *fenestra vestibuli* (oval window)	4 mm
Length of the basilar membrane	3.35 mm
Width of the basilar membrane at its beginning	0.04 mm
Width of the basilar membrane at its end	0.495 mm
Number of hair cells on the basilar membrane	14,000
Number of coils in the cochlea	2.5
Spatial resolution of the ear	8.4°

Table 4.6.10: Age Dependency of Upper Hearing Limits

The most important pitches (frequencies) of human language lie between ca. 80 and 600 Hz; consonants of the S and T groups lie between ca. 1 and 3 kHz. For animals, see Table 1.11.13.

Figures from REIN and SCHNEIDER 1971

Age	Upper limit (kHz)
Child	21
35 years	15
59 years	12
Elderly man	5

Table 4.6.11: Sound Levels of Selected Sounds

Figures given in A-weighted decibels [dB(A)]. dB(A) describes the intensity of the frequency-dependent perception of the human ear. At a frequency of 1,000 Hz, dB(A) values coincide with phon values. Since decibel [dB(a)] values represent logarithmic units, a sound that increases by 10 dB(A) is sensed to be twice as loud.

Figures after BELTZ 1974, FAHRENBERGER and MÜLLER 1972, REIN and SCHNEIDER 1971

	dB(A)
Hearing threshold	0
Quiet whispering, ticking of a pocket watch, breathing in an echo-free space	10
Whispered dialogue (from a film recording studio)	20
Alarm clock, residential area at night	30

(Cont.) →

	dB(A)
Quiet conversation, quiet apartment	40
Normal conversation, loudspeaker set at room volume	50
Conversation	60
Office typewriter	70
Vacuum cleaner, road traffic	80
Motorcycle, automobile, heavy road traffic	90
Jackhammer, a nearby car horn	100
Textile factory	110
Propeller airplane, rock 'n' roll band	120
Pain threshold, noise in a boiler shop	130
Jet fighter at takeoff	140
Jet engine with afterburners at 25 meters	160
Space rocket at blastoff, heard from nearby	170

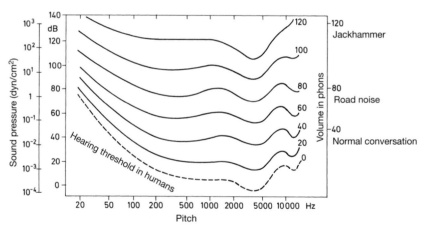

Fig. 4.6.11: Relationship between the objectively measurable, frequency-independent sound pressure (dB) and the subjectively perceived, frequency-dependent sound level (phon). From REMANE, STORCH and WELSH 1974

Table 4.6.12: Spatial Discrimination Threshold of Selected Sites on the Human Body

The spatial discrimination threshold (or spatial resolution) represents the distance separating two equally strong pressure stimuli (e.g., with a pair of compasses) at which they can still be perceived as two separate stimuli.

From CZIHAK et al. 1976, MEYER 1964, OPPENHEIMER and PINCUSSEN 1925

Body site	Spatial threshold (mm)	Body site	Spatial threshold (mm)
Tip of the tongue	1.1	Buttocks	45.0
Border of the tongue	9.0	Fingertip	2.3
Lip	4.0	Back of the finger	6.8
Hard palate	13.6	Palm of the hand	11.3
Tip of the nose	6.8	Back of the hand	31.6
Cheek	11.3	Forearm, underside	15.0
Eyelid	11.3	Forearm, back	40.0
Forehead	22.6	Upper arm	68.0
Chin	33.8	Kneecap	36.0
Sternum	45.0	Thigh	68.0
Nape of the neck	67.7	Lower leg	40.6
Back	68.0	Heel	22.6

Table 4.6.13: Number and Distribution of Hot and Cold Spots in Human Skin

nr = not resolvable into spots.

From GAUER et al. 1972

Body region	Number of cold spots per cm^2	Number of hot spots per cm^2
Forehead	5.5– 8	2
Nose	8 –13	1
Mouth	16 –19	nr
Rest of face	8.5– 9	1.7
Breast	9 –10.2	0.3
Forearm	6 – 7.5	0.3–0.4
Palms of the hands	1 – 5	0.4
Backs of fingers	7 – 9	1.7
Finger, underside	2 – 4	1.6
Thigh	4.5– 5.2	0.4

Table 4.6.14: Odor Threshold in Humans for Selected Distinctive Substances

Data after GANONG 1972, KEIDEL 1971, *Lexikon der Biologie*, REIN and SCHNEIDER 1971

Substance	Olfactory threshold (mg/l air)
Pyrazine	300
Ethanol	100
Ethyl ether	5.83
Chloroform	3.30
Naphthalene	0.04
Pyridine	0.03
Peppermint oil	0.02
Iodoform	0.02
Propyl mercaptan	0.006
Acetic acid	0.001
Vanilla	0.00005
Synthetic musk	0.00004
Butyric acid	0.00001
Skatole	0.0000035
α-Ionone	0.0000005
Methyl mercaptan	0.0000004

Table 4.6.15: Taste Thresholds in Humans

Figures represent the mimimum amount of the substance in question in grams per liter of water.

Figures from KEIDEL 1971

Taste quality	Substance	Threshold (g/l)
Sweet	Saccharin	0.001
Sour	Hydrochloric acid	0.01
Salty	Common salt	1.0
Bitter	Quinine	0.004

Part 5: Comparative Data

Table 5.1: Composition of Animal and Plant Organisms from the Elements

Figures represent the percentage of total weight. Please note that just the four elements of carbon, hydrogen, oxygen, and nitrogen account for 99.77% of the weight of plants and 96.54 % of animals. See also Table 4.1.3, for the specific figures on the human organism.

From HEIDERMANNS 1957, modified

	Animal (here: mammal)	Plant (here: stone pine)
Carbon (C)	21.15	53.960
Hydrogen (H)	9.86	7.130
Oxygen (O)	62.43	38.650
Nitrogen (N)	3.10	0.030
Calcium (Ca)	1.90	0.007
Phosphorus (P)	0.95	0.005
Potassium (K)	0.23	0.006
Sulphur (S)	0.16	0.052
Chlorine (Cl)	0.08	0.001
Magnesium (Mg)	0.027	0.003
Iodine (I)	0.002	0.002
Iron (Fe)	0.005	0.030
Fluorine (F)	0.003	0.001
Aluminum (Al)	0.001	0.065
Silicon (Si)	0.001	0.057
Manganese (Mn)	0.0005	0.001

Table 5.2: Water Content of Organisms

Figures as percentage of total weight. See also Tables 1.1.19 and 2.1.10.

Data after BUDDENBROCK 1956, HEIDERMANNS 1957, STRASBURGER 1978

Species	Water content (%)
Plants	
Algae	up to 98.0
Slime molds	70–94.0
Leaves. higher-order plants	80–90.0
Cucumber	up to 95.0
Wood	50.0
Dry seeds	13–14.0
Animals	
Moon jelly	98.2
Leech (*Herpobdella*)	86.3
Ascaris worm	79.3
Edible snail (without shell)	84.4

(Cont.) →

Species	Water content (%)
Terrestrial nematodes (*Planorbis*)	88.3
Water flea	73.9
Caterpillar of the small white (butterfly)	83.0
Colorado potato beetle	62–66.0
Grain weevil	40–44.0
Tench	80.0
Frog	77.0
Sheep	57.0
Pig	55.0
Rabbit	69.0
Human	
Adult	60.0
Embryo, 15 weeks	91.5
Fetus, 24 weeks	85.0
Fetus, 30 weeks	79.0
Newborn	66.0

Table 5.3: Size and Molecular Weight of Selected Important Biological Molecules

Data after CZIHAK et al. 1976, HARTMANN 1953, HEIDERMANNS 1957, KARLSON 1974, MEYER 1964

	Size		Mol. wt.
Myoglobin	0.07	μm	17,500
Myosin			468,000
Fibrinogen	0.68	μm	400,000
Serum albumin (horse)	0.28	μm	66,900
Serum globulin (horse)	0.56	μm	150,000
Pepsin	0.14	μm	35,500
Urease	1.68	μm	483,000
Insulin	0.14	μm	35,500
Catalase	1.12	μm	281,600
Hemoglobin (human)	2.8	μm	66,800
Hemocyanin	up to 26.88	μm	up to 6,300,000
Glycogen			$1–16 \times 10^6$
Cellulose	8.3	μm	2,400,000
Glucose	0.7	nm	180
Chlorophyll	3.5	nm	approx. 900
Amino acid glycine	7	Å	75
Amino acid leucine	12	Å	131

Table 5.4: Redox Potentials of Selected Organic Compounds

Since the redox potential depends upon the given pH value, the latter is also indicated.

After *Enzyklopädie* 1979, Hirsch 1966

	pH value	Potential E_0 (volt)
Cytochrome a	7.4	+ 0.29
Cytochrome c	5–8	+ 0.26
Cytochrome b	7.4	– 0.04
Hemoglobin	7.0	+ 0.152
Ascorbic acid	4.59	+ 0.136
Methylene blue	7.0	+ 0.011
Pyocyanin	7.0	– 0.034
Flavoprotein	7.0	– 0.059
Hemin	6.99	– 0.114
Lactate/pyruvate	7.01	– 0.186
Riboflavin	7.0	– 0.208
Cystine/cysteine	7.15	– 0.227
Gluthathione	7.15	– 0.233
Coenzyme I	7.4	– 0.325

Table 5.5: Dimensions of Cells and Organelles

Data after Czihak et al. 1976, David 1977, Metzner 1966, Meyer 1964, Scharf and Weber 1976, Strasburger 1978

	Length		Diameter	
Foot-and-mouth disease virus			10	nm
Influenza virus			120	nm
Tobacco mosaic virus	28	nm		
Micrococcus (bacterium)			0.2	μm
Escherichia coli	3	μm		
Thiospirillum (bacterium)	80	μm		
Yeast cells			6– 8	μm
Chlamydomonas (flagellate alga)			20	μm
Diatoms, pennate	180	μm		
Diatoms, centric			35	μm
Red blood cells, human			7.5	μm
Spermatozoon, human	50	μm	2	μm
Egg cell, human			100	μm
Ostrich egg cell	150	mm		
Epidermis, oak			28	μm
Pith, elder			200	μm
Epidermis, onion			400	μm

(Cont.) →

	Length	Diameter
Fiber cell, flax	40–65 mm	
Fiber cell, nettle	50–75 mm	
Internodal cell, *Chara*	40–80 mm	
Fiber cell, ramie	400 mm	
Oral mucosa cell, human		60–80 μm
Mitochondria		0.5– 0.8 μm
Dictyosomes	0.2– 5.5 μm	
Chloroplasts		4– 8 μm
Ribosomes		10–15 nm
Simple membrane, plants		6– 8 nm
Double membrane, endoplasmatic reticulum, plants		25–30 nm
Cell membrane, human, oral mucosa		7.67 nm
Cell membrane, human, erythrocyte		8.5 nm

Table 5.6: Speed of Movement of Cells and Organelles

Figures from CZIHAK et al. 1976, *Documenta Geigy* 1975, HERTEL 1963, KÄSTNER 1973,
OPPENHEIMER and PINCUSSEN 1927, PROSSER 1973

Cell or organelle	Speed (μm/s)
Bacillus subtilis	10
Bacillus megatherium	200
Euglena	130– 200
Amoeba (*A. limax*)	1
Diatom	7
Paramecium	2,000–3,000
Zoospore from sea lettuce	150
Spermatozoids of the liverwort *Marchantia*	20– 27
Migrating plasmodia	14– 17
Spermatozoa, cattle	200
Spermatozoa, human	1,300–2,600
White blood cells, human	4– 24
Lymphocytes in tissue culture	0.55
Chromosomes during cell division	0.01– 0.03
Golgi vesicle (plant cell)	0.1– 0.3
Vesicle transport down the neuronal axon	1– 10
Plasma flow in root hairs	2– 16
Plasma flow in *Nitella*	80–1,000

Table 5.7: Times of Cell Division in Selected Cells

Data after Berkaloff et al. 1973, Hirsch 1966, Kuhl 1964, Nougarède 1969, Oppenheimer and Pincussen 1930

Species and cell	Prophase	Metaphase	Anaphase	Telophase	Overall time
Heliozoa	12 min	3–6 min	4 min		30 min
Broad bean, meristem	90 min	31 min	34 min	34 min	189 min
Pea, endosperm	40 min	20 min	12 min	110 min	182 min
Iris, endosperm	40–60 min	10–30 min	12–22 min	40–75 min	140 min
Grasses, stomata cells	36–45 min	7–10 min	15–20 min	20–35 min	78–110 min
Newt, fibroblasts	18 min	38 min	26 min	28 min	110 min
Chicken, fibroblasts	45 min	6 min	2 min	10 min	63 min
Chicken, mesenchyme	30–60 min	2–10 min	2–3 min	3–12 min	37–82 min
Rat, hepatocytes	4 h	10 min	30 min	30 min	5 h 10 min
Sea urchin eggs, furrow formation					30–40 min
Drosophila eggs, furrow formation	4 min	30 s	1 min	50 s	6 min 20 s

Table 5.8: Dimensions of Cell Nuclei

Figures for selected organs and animals

Figures from David 1977

Species	Organ	Diameter (µm)	Volume (µm^3)
Mouse	Liver	10.1	528
	Pancreas	6.4	142
	Cardiac muscle	8.8 × 2.3	
	Striated muscle	10.4 × 3.3	
	Smooth muscle	10.2 × 2.2	
Rat	Adrenal cortex	7.1	187
Guinea pig	Spinal ganglion	12.0	915
Rabbit	Spinal ganglion	14.8	1,678
Cat	Spinal ganglion	13.3	1,245
Dog	Spinal ganglion	13.9	1,415
Cattle	Spinal ganglion	16.5	2,338

(Cont.) →

Species	Organ	Diameter (μm)	Volume (μm³)
Human	Kidney	6.2 × 5.5	
	Spinal ganglion	14.2	1,509
	Lung. alveoli	5.8 × 3.7	
Amphibians	Oocyte	480	

Table 5.9: Dimensions of the Mitochondria of Selected Organs and Organisms

After KLIMA 1967

Species, Organ	Diameter (μm)	Length (μm)
Euglena	0.50	10.0
Bread mold (*Neurospora*)	0.50	3.0
Oats, coleoptiles	0.50	1.0
Migratory locust. wing muscle	1.00	8.0
Blowfly, wing muscle	1.00	3.0
Alpine newt, chorda	0.3 –0.50	3.0
Sparrow, cardiac muscle	1.00	2.0
Rat, cardiac muscle	0.4 –0.70	1.5
Rat, kidney, proximal tube	0.5 –0.70	8.0
Rat, spinal marrow neuron	0.20	2.5
Guinea pig, ovary	0.5 –0.75	1.0
Golden or Syrian hamster, small intestine	0.2 –0.25	2.5
Human, small intestine	0.5 –1.20	2.0
Human, retinal rods	0.18–0.25	4.0

Table 5.10: Mitochondria: Cell Share

Figures represent mitochondria volume as a proportion of total cell volume, given as percentage.

Figures from DAVID 1977

Species	Organ	Mitochondria (%)
Housefly	Wing muscle	28.2
Migratory locust	Wing muscle	30
Sticklebacks (gasterosteidae)	Pituitary gland	4.2
Cod (Baltic Sea)	Red muscle fiber	25.39
	White muscle fiber	1.17

Species	Organ	Mitochondria (%)
Frog	Liver	5.38
Rock dove	Liver	20.02
Hedgehog	Spinal ganglion	14.7
Mouse	Liver	16.93
	Pituitary gland	7.5
	Heart, ventricle	47.7
	Heart, atrium	34.9
	Leukocytes	3.6
Rat	Liver	18.38
	Pancreas	7.9
	Kidney	37.23
	Heart	35.8
Guinea pig	Smooth muscle cell	4.7
Rabbit	Liver	11.46
Cattle	Liver	17.75
Human	Liver	13.22
	Small intestine	13.34

Table 5.11: DNA Content of Diploid Cell Nuclei

Figures represent the amount of DNA per cell nucleus in picograms. For the chromosome sets of selected animals, see Table 1.9.3; for plants, see Table 2.2.9.
Data after BERKALOFF et al. 1973, BRESCH and HAUSMANN 1972, NAGL 1976

Organism	DNA per nucleus (pg)
Escherichia coli (bacterium)	0.0042
Pine	93.8
Fir	100.0
Wheat	36.2
Rye	17.7– 18.9
Corn (maize)	6.6– 15.4
Barley	12.8
Meadow grass (*Poa annua*)	7.0
Onion	33.5
Tradescantie	116.0
Lesser celandine	19.6
Tobacco	13.0
Pea	9.1
Shepherd's purse	1.7
Sunflower	9.8
House cricket	12.0
Fruit fly (*Drosophila*)	0.2

(Cont.) →

Organism	DNA per nucleus (pg)
Goldfish	4.0
Carp	3.4
Pike	1.7
Trout	4.9
Fire salamander	85.3
Edible frog	16.8
Common toad	15.0
Tree frog	11.6
Turtle	5.3
Snakes	5.0
Chicken	2.5
Goose	2.9
Sparrow	1.9
Rat	5.7
Mouse	5.0
Guinea pig	5.9
Dog	5.3
Cattle	6.4
Sheep	5.7
Human	5.8

Table 5.12: Size of DNA in Selected Organisms

For organisms with chromosomes, figures apply to the haploid chromosome set. For absolute quantity in picograms, see Table 5.11.

Figures from *Lexikon der Biologie*

Species	Length of DNA		Base pairs	
Mitochondrial DNA, human	5.6	μm	16,569	
Mitochondrial DNA, yeast	25	μm	7.5	$\times 10^3$
Chloroplast DNA corn (maize)	44	μm	140	$\times 10^3$
Bacteriophage λ	16.5	μm	48,502	
Escherichia coli	1.36	mm	4	$\times 10^6$
Yeast (*Saccharomyces cerevisiae*)	4.6	mm	13.5	$\times 10^6$
Fruit fly (*Drosophila melanogaster*)	56	mm	160	$\times 10^6$
Sea urchin (*Strongylocentrotus purpuratus*)	170	mm	800	$\times 10^6$
African clawed frog (*Xenopus laevis*)	1	m	3	$\times 10^9$
Mammals, including humans	1	m	3	$\times 10^9$

Table 5.13: Base Ratios in the DNA of Selected Organisms

A = adenine, C = cytosine, G = guanine, T = thymine.
Note that there are equal quantities of each of the respective bases within a given complimentary pair (thus equal quantities of adenine and thymine, on the one hand, and of guanine and cytosine, on the other).

Figures from BRESCH and HAUSMANN 1972, CZIHAK, LANGER and ZIEGLER 1976

Species	(A + T) : (G + C)
Viruses	
Herpes simplex virus	0.47
T2 phage	1.86
Bacteriophage λ	0.79
Bacteria	
Sarcina lutea	0.35
Mycobacterium tuberculosis	0.60
Escherichia coli	0.93–0.97
Vibrio cholerae	1.31
Bacillus subtilis	1.66
Plants and Fungi	
Yeast (*S. cerevisiae*)	1.79
Black mold (*Asp. niger*)	1.00
Euglena gracilis, cell nucleus	0.88
Euglena gracilis, plastid DNA	3.23
Wheat	1.19–1.22
Animals	
Migratory locust	1.41
Sea urchin (*Paracentrotus*)	1.84
Sea urchin (*Arbacia*)	1.58
Crab (*Cancer borealis*)	17.50
Trout	1.34
Herring	1.23
Chicken	1.38
Sheep	1.36
Cattle	1.30
Horse	1.33
Human	1.40–1.52

Table 5.14: Differences in the Number of Amino Acids in Humans Versus Selected Organisms as Exemplified by Cytochrome c.

Cytochromes are among the oldest of proteins and are made up of somewhat more than 100 amino acids. For this reason, differences in amino acids allow us to draw conclusions about the degree of relatedness or the period of divergent development of the organisms being compared.

Figures from DICKERSEN and GEIS 1971

Organisms compared	Number of differing amino acids in cytochrome c
Human – rhesus monkey	1
Human – rabbit	9
Human – gray whale	10
Human – cow, sheep, pig	10
Human – kangaroo	10
Human – donkey	11
Human – dog	11
Human – Peking duck	11
Human – horse	12
Human – chicken, turkey	13
Human – penguin	13
Human – rattlesnake	14
Human – snapping turtle	15
Human – bullfrog	18
Human – tuna	21
Human – fly (*Chrysomyia*)	27
Human – silkworm	31
Human – wheat	43
Human – baker's yeast	45
Human – sac fungus (*Neurospora crassa*)	48
Human – yeast (*Candida krusei*)	51

References

Abderhalden, R.: Die Wirkstoffe: Vitamine, Hormone und Fermente. In: Bertalanffy, L.v. (ed.): Handbuch der Biologie, vol. II, Frankfurt 1966

Altman, P.L. and D.S. Dittmer (eds.): Environmental Biology. Bethesda 1966

Altman, P.L. and D.S. Dittmer (eds.): Respiration and Circulation. Bethesda 1971

Altman, P.L. and D.S. Dittmer (eds.): Biology Data Book. vols. I – III, 2. edn., Bethesda 1972–1974

Amann, G.: Kerfe des Waldes. 7. edn., Melsungen 1976

Andreae, B.: Räumliche Grenzen des Nahrungsspielraums, Naturw. Rdsch. 29, 11, 1976

Aurich, H.: Laboratorium des Lebens. Cologne 1978

Austin, C.R.: Die Befruchtung. Stuttgart 1967

Austin, C.R. and R.V. Short: Fortpflanzungsbiologie der Säugetiere. Berlin 1978

Autrum, H.: Über Energie- und Zeitgrenzen der Sinnesempfindungen. Die Naturw. 35, 361, 369, 1948

Bassus, W.: Gifte im Tierreich. Die Neue Brehm Bücherei 354, Wittenberg 1965

Bauer, H.C., Hofer, R., Knapp, W. and H. Moser: Zoologische Experimente. Munich 1974

Bauer, R.D., Pasch, Th. and E. Wetterer: Biomechanik des Blutkreislaufs. In: Hoppe et al.: Biophysik. Heidelberg 1977

Bechtel, H., Cribb, P. and E. Launert: Orchideenatlas. Stuttgart 1980

Bellairs, A.: Die Reptilien. In: Die Enzyklopädie der Natur, vols. 10, 11. Lausanne 1971

Berkaloff, A., Bourguet, J., Favard, P. and M. Guinnebault: Biologie und Physiologie der Zelle. Braunschweig 1973

Berndt, R. and W. Meise: Naturgeschichte der Vögel. Stuttgart 1959

Bertelsmann Lexikon-Institut: Biologie. Daten und Fakten zum Nachschlagen. Gütersloh 1979

Beltz-Praxis: Versuche zum Umweltschutz. 2. edn., Weinheim and Basel 1974

Blinker: Die Blinker-Hitparade, Blinker 11, 38 – 39, 2001

Bogner, H. and H.-Chr. Ritter (eds.): Tierproduktion. Ulmers Taschenbücher. Stuttgart 1976

Brauner, L.: Einführung in die Physikalische Chemie der Pflanzenzelle. In: Bertalanffy, L v. (ed.): Handbuch der Biologie I/1. Konstanz 1965

Bresch, C. and R. Hausmann: Klassische und molekulare Genetik. 3. edn., Berlin 1972

Brink, F.H. van den: Die Säugetiere Europas. Hamburg 1975

Brohmer, P.: Fauna von Deutschland. 14. edn., Heidelberg 1977

Brouwer, W.: Handbuch des speziellen Pflanzenbaus. Berlin and Hamburg 1972

Bruggaier, W. and D. Kallus: Biologie des Menschen. Frankfurt 1973

Buddenbrock, W.v.: Grundriß der vergleichenden Physiologie. vols. I – IV., Berlin 1924, Basel 1956 – 1967

Buddenbrock, W.v.: Wie orientieren sich die Tiere. Stuttgart 1956

Bünning, E.: Entwicklungs- und Bewegungsphysiologie der Pflanze. Berlin, Göttingen, Heidelberg 1953

Cochran, D.M.: Knaurs Tierreich in Farben: Amphibien. Munich 1970

Creutz, G.: Geheimnisse des Vogelzugs. 5. edn. Die Neue Brehm Bücherei 75, Wittenberg 1965

Csicsáky, M.: Vogel und Flugzeug: Wie weit geht die Analogie? BIUZ 14/2, 48–52, 1984

Czihak, G., Langer, H. and H. Ziegler (ed.): Biologie. Berlin, Heidelberg, New York 1976

David, H.: Quantitative Ultrastructural Data of Animal and Human Cells. Stuttgart, New York 1977

Deutscher Jagdschutz-Verband: DJV-Handbuch 1978. Mainz 1978

Dickersen, R.E. and I. Geis: Struktur und Funktion der Proteine. Weinheim 1971

Die Roten Listen (ed. E. Jedicke). Stuttgart 1997

Dittmer, D.S. and R.M. Grebe (eds.): Handbook of Respiration. Philadelphia and London 1958

Documenta Geigy: Wissenschaftliche Tabellen. Stuttgart 1975

Dorst, J.: Das Leben der Vögel. In: Die Enzyklopädie der Natur, vols. 12, 13. Lausanne 1972

Dorst, J. and P. Dandelot: Säugetiere Afrikas. Hamburg and Berlin 1973

Ebel, F. and O. Birnbaum: Schöne und seltsame Welt der Orchideen. Tübingen 1971

Ellenberg, H.: Vegetation Mitteleuropas mit den Alpen in ökologischer Sicht. 2. edn. Stuttgart 1978

Fahrenberger, G. and J. Müller: Luft und Wasser in Gefahr. Göttingen 1972

Farmer, J.B.: On the Quantitative Differences in the Water-Conductivity of the Wood in Trees and Shrubs. Parts I und II. Proc. Roy. Soc., B, 90, 218–250, 1918

Flindt, R.: Ökologie im Jahreslauf. Biologische Arbeitsbücher 2. 4. edn. Heidelberg 1985

Florey, E.: Lehrbuch der Tierphysiologie. Stuttgart 1970

Francois, J. and F. Hollwich: Augenheilkunde in Klinik und Praxis. Stuttgart 1977

Frank, W.: Parasitologie. Stuttgart 1976

Franke, W.: Naturpflanzenkunde. Stuttgart 1976

Franz, E.: Der Flug der Insekten. Natur und Volk 89, 1959

Ganong, W.F.: Medizinische Physiologie. 2. edn. Berlin, Heidelberg 1972

Gauer, O.H., Kramer, K. and R. Jung (eds.): Physiologie des Menschen. Munich 1971

George, U.: Flugechsen, die Saurier der Lüfte. Geo 11, 14, 1994

Geßner, F.: Die Leistungen des pflanzlichen Organismus. In: Bertalanffy, L.v. (ed.): Handbuch der Biologie vol. IV. Potsdam 1942

Giersberg, H. and P. Rietschel: Vergleichende Anatomie der Wirbeltiere. Jena 1968

Goadby, P.: Billfishing. Camden 1996

Götz, A. and J. Konrad: Pflanzenbau. Landwirtschaftliches Lehrbuch I. 5. edn. Stuttgart 1977

Graf, J.: Tafelwerk zur Pflanzensystematik. Munich 1975

Grafe, A.: Viren. Parasiten unseres Lebensraumes. Heidelberger Taschenbücher 192. Berlin, Heidelberg, New York 1977

Greulach, V.A. and J.E. Adams: Plants – An Introduction to Modern Botany. New York, London 1962

Grimme, L.H.: Biologische und biotechnologische Beiträge zur Welternährungslage. Naturw. Rdsch. 28, 5, 1975

Grzimek, B. (ed.): Grzimeks Tierleben. Zürich 1970

Guttenberg, H.v.: Lehrbuch der allgemeinen Botanik. 6. edn. Berlin 1963

Haas, J.: Der Ursprung des Lebens. Munich 1964

Hager, P.: Vom Leben des Ackers. Stuttgart 1960

Halstead, L.B.: Die Welt der Dinosaurier. Hamburg 1975

Haltenorth, T.: Die Stämme des Tierreichs. III. In: Bertalanffy, L.v. and F. Geßner (eds.): Handbuch der Biologie, vol. VI/3. Wiesbaden 1977

Haltenorth, T.: Rassehunde – Wildhunde. Winters Naturw. Taschenb. Heidelberg 1958

Haltenorth, T. and W. Trense: Das Großwild der Erde und seine Trophäen. Bonn, Munich, Vienna 1956

Hanf, M.: Die Ackerunkräuter und ihre Keimlinge. Limburgerhof (no date)

Hanke, W., Hamdorf, K., Horn, E. and C. Schlieper: Praktikum der Zoophysiologie. Stuttgart, New York 1977

Harms, J.W.: Der Lebenslauf der Tiere. In: Bertalanffy, J.v. (ed.): Handbuch der Biologie vol. II. Frankfurt 1966

Harrison, C.: Jungvögel, Eier und Nester aller Vögel Europas, Nordafrikas und des Mittleren Ostens. Hamburg 1975

Heberer, G.: Homo – Unsere Ab- und Zukunft. Stuttgart 1968

Hegi, G.: Illustrierte Flora von Mittel-Europa. 2. edn. Munich 1965

Heidermanns, C.: Grundzüge der Tierphysiologie. 2. edn. Stuttgart 1957

Heinroth, O.: Aus dem Leben der Vögel. Verst. Wiss. 34, Berlin 1955

Heller, R.: Biologie végétale. vol. II: Nutrition et métabolisme. Paris 1969

Hertel, H.: Struktur, Form, Bewegung. Mainz 1963

Herter, K.: Der Temperatursinn der Tiere. Die Neue Brehm Bücherei 295, Wittenberg 1962

Herzog, K.: Anatomie und Flugbiologie der Vögel. Stuttgart 1968

Hess, H.E., Landolt, E. and R. Hirzel: Flora der Schweiz. Basel and Stuttgart 1967

Heynert, H.: Grundlagen der Bionik. Heidelberg 1976

Heywood, V.H. (ed.): Flowering Plants of the World. Oxford 1978

Higgins, L.G. and N.D. Riley: Die Tagfalter Europas und Nordwestafrikas. Hamburg and Berlin 1971

Hirsch, G.C.: Dynamik der tierischen Zelle. In: Bertalanffy, L.v. (ed.): Handbuch der Biologie, vol. I/1. Konstanz 1965

Hirsch, G.C.: Die Zellorganellen und ihre Zusammenarbeit. In: Bertalanffy, L.v. (ed.): Handbuch der Biologie, vol. I/2. Frankfurt 1966

Hirsch, G.C.: Die Lebensäußerungen der Tiere. In: Bertalanffy, L.v. (ed.): Handbuch der Biologie, vol. V. Wiesbaden 1977

Hoeher, S.: Gelege der Vögel Mitteleuropas. Melsungen 1972

Hofmeister, H.: Lebensraum Wald. Munich 1977

Jacobs, W. and M. Renner: Taschenlexikon zur Biologie der Insekten. Stuttgart 1974

Jensen, W.A. and F.B. Salisbury: Botany: An Ecological Approach. Belmont 1972

Kästner, A.: Lehrbuch der Speziellen Zoologie. Stuttgart 1973

Karcher, H.L.: Wie ein Ei dem anderen. Munich and Zürich 1975

Karlson, P.: Kurzes Lehrbuch der Biochemie. 11. edn. Stuttgart 1980

Katz, B.: Nerv, Muskel und Synapse. Stuttgart 1971

Kaussmann, B.: Pflanzenanatomie. Jena 1963

Keidel, W.D.: Kurzgefaßtes Lehrbuch der Physiologie. Stuttgart 1970

Kemper, H. and E. Döhring: Die sozialen Faltenwespen Mitteleuropas. Berlin and Hamburg 1967

Kerner v. Marilaun, A.: Pflanzenleben. vols. I and II., 2. edn. Leipzig and Vienna 1898

Kleiber, M.: Der Energiehaushalt von Mensch und Haustier. Hamburg 1967

Klima, J.: Cytologie. Stuttgart 1967

Koenigswald, G.v.: Die Geschichte des Menschen. Verst. Wiss. 74, Berlin 1960

Kolb, E. (ed.): Lehrbuch der Physiologie der Haustiere. 3. edn. Stuttgart 1974

Korschelt, E.: Lebensdauer, Altern und Tod. 3. edn. Jena 1924

Kosmos-Lexikon der Naturwissenschaften. Stuttgart 1953

Kraus, O. and K. Kubitzky: Biologische Systematik. Denkschrift der DFG. Weinheim 1982

Kremer, B.P.: Stoffwechselstrategie und Produktivität der großen Meeresalgen. Naturw. Rdsch. 33, 4, 1980

Krumbiegel, J.: Von Haustieren und ihrer Geschichte. Stuttgart 1947

Krumbiegel, J.: Biologie der Säugetiere. Krefeld 1955

Krumbiegel, J.: Einhufer. Die Neue Brehm Bücherei 208, Wittenberg 1958

Kuhn-Schnyder, E.: Die Saurier der Trias. Stuttgarter Beitr. zur Naturkunde, Series C, No. 1, Stuttgart 1974

Kuhn, O.: Lurche und Kriechtiere der Vorzeit. Die Neue Brehm Bücherei 217, Wittenberg 1958

Kunstyr, I., Schiel, R., Kaup, F.J., Uhr, G. and H. Kirchhoff: Giant Gram-Negative Noncultivable Endospore-Forming Bacteria in Rodent Intestines. Naturwissenschaften 75, 525–527, 1988

Ladiges, W. and D. Vogt: Die Süßwasserfische Europas. Hamburg 1965

Laget, P.: Biologie et physiologie des éléments nerveux. Paris 1970

Lampeitl, F.: Bienen halten, eine Einführung in die Imkerei. Munich 1982

Langmann, J.: Medizinische Embryologie. Stuttgart 1976

Larcher, W.: Ökologie der Pflanzen. 2. edn. UTB 232, Stuttgart 1976

Laskin, A.J. and H.A. Lechevalier (eds.): Handbook of Microbiology. Cleveland 1973

Lehmann, K.: Zum Alter von Testudo (Asterochelys) radiata in Gefangenschaft II (Testudines: Testudinidae). Salamandra 22, 281–282, 1986

Lehnartz, E.: Die chemischen Voraussetzungen des Lebens. In: Bertalanffy, L.v. (ed.): Handbuch der Biologie, vol. I/1. Konstanz 1965

Leukeit, W., Breirem, K. and E. Crasemann (eds.): Handbuch der Tierernährung. Hamburg 1969

Lexikon der Biologie. Freiburg 1983 – 87

Libbert, E.: Lehrbuch der Pflanzenphysiologie. 3. edn. Stuttgart 1979

Lötschert, W.: Pflanzen an Grenzstandorten. Stuttgart 1969

Lofts, B. (ed.): Physiology of the Amphibia. New York 1976

Lüttschwager, J.: Die Drontevögel. Die Neue Brehm Bücherei 276, Wittenberg 1961

Lundegardh, H.: Pflanzenphysiologie. Jena 1960

Madle, H.: Lehrbuch der Naturwissenschaften. vol. 1: Biologie. Stuttgart 1967

Mann, D.: Rekorde der Tierwelt. Munich and Berlin 1979

Mauch, H. and K. Lauber: Unsere Pilze. Hallwag Taschenbuch 10, 2. edn. Bern 1975

McGregor, P. and H. Edwards: Reitpferde. Wiesbaden 1976

Mebs, D.: Hauttoxine bei Farbfröschen. Naturw. Rdsch. 32, 7, 1979

Mengel, K.: Ernährung und Stoffwechsel der Pflanze. Stuttgart 1968

Merck: Labormethoden des praktischen Arztes. Darmstadt 1966

Mertens, R.: Die Lurche und Kriechtiere des Rhein-Main-Gebietes. Frankfurt 1947

Metzner, H. (ed.): Die Zelle, Struktur und Funktion. Stuttgart 1966

Meyers Handbuch über Mensch, Tier und Pflanze. Mannheim 1964

Michael, E., Hennig, B. and H. Kreisel: Handbuch für Pilzfreunde. 3. edn. Jena 1978

Michaelis, H.J.: Der Wellensittich. Die Neue Brehm Bücherei 244, Wittenberg 1974

Michel, G.: Kompendium der Embryologie der Haustiere. 4. edn. Stuttgart 1986

Mitchell, H.H., Hamilton, T.S., Steggerda, F.R. and H.W. Bean: Chemical Composition of the Adult Human Body and its Bearing on the Biochemistry of Growth. J. Biol. Chem. 158, 625 – 637, 1945

Molisch, H.: Botanische Versuche ohne Apparate. 4. edn. Stuttgart 1965

Mossmann, D.J. and W.A.S. Sarjeant: Fossile Fährten – den Dinosauriern auf der Spur. Spektrum der Wiss. 1983/3, 68 – 79

Mühlenberg, M.: Freilandökologie. UTB 595, Heidelberg 1976

Müller-Schneider, P.: Verbreitungsbiologie der Blütenpflanzen. Veröff. Geobot. Inst. ETH Zürich 1977

Nachtigall, W.: Biostatik. In: Hoppe et al.: Biophysik. Heidelberg 1977

Nachtigall, W.: Unbekannte Umwelt. Hamburg 1979

Nagl, W.: Zellkern und Zellzyklen. Stuttgart 1976

Niethammer, J.: Säugetiere. UTB 732, Stuttgart 1979

Nigon, V. and W. Lueken: Vererbung. Allgemeine Biologie, vol. 4. Stuttgart 1976

Nilsson, S., Praglowski, J. and L. Nilsson: Atlas of Airborne Pollen Grains and Spores in Northern Europe. Stockholm 1977

Norman, J.R. and F.C. Fraser: Riesenfische, Wale und Delphine. Hamburg 1963

Nüßlein, F.: Jagdkunde. Munich 1977

Oppenheimer, C. and L. Pincussen: Tabulae biologicae. vols. I – V. Berlin 1925 – 1927

Otte, H.J. and H. Brandis: Lehrbuch der Medizinischen Mikrobiologie. 4. edn. Stuttgart 1978

Penzlin, H.: Lehrbuch der Tierphysiologie. 3. edn. Stuttgart 1981

Perrey, F.: Blumen der Welt. Freiburg 1972

Pflugfelder, O.: Lehrbuch der Entwicklungsgeschichte und Entwicklungsphysiologie der Tiere. Jena 1962

Pisek, A. and E. Berger: Kutikuläre Transpiration und Trockenresistenz isolierter Blätter und Sprosse. Planta 28, 124 – 155, 1938

Platt, D.: Biologie des Alterns. Heidelberg 1976

Portmann, A.: Die Stellung des Menschen in der Natur. In: Bertalanffy, L.v. (ed.): Handbuch der Biologie IX, Konstanz 1965

Portmann, A.: Einführung in die vergleichende Morphologie der Wirbeltiere. Basel and Stuttgart 1959

Precht, H., Christophersen, J. and H. Hensel: Temperatur und Leben. Berlin 1955

Prosser, C.L. (ed.): Comparative Animal Physiology. 3. edn. Philadelphia, London, Toronto 1973

Rahmann, H.: Die Sehschärfe bei Wirbeltieren. Naturw. Rdsch. 20, 1, 8 – 14, 1967

Rein, H. and M. Schneider: Einführung in die Physiologie des Menschen. 16. edn. Berlin 1971

Remane, A., Storch, V. and U. Welsch: Kurzes Lehrbuch der Zoologie. 5. edn. Stuttgart 1985

Riesner, D.: Viroide – die kleinsten infektiösen Krankheitserreger. Med. i. and Z. 3, 158 – 165, 1979

Rohrlich, M. and G. Brückner: Das Getreide. 2. edn. Berlin and Hamburg 1967

rororo-Pflanzenlexikon. Reinbeck 1976/77

Rucker, E.: Der Mensch in Zahlen. Munich 1967

Saller, K.: Leitfaden der Anthropologie. 2. edn. Stuttgart 1964

Sanderson, I.T.: Säugetiere. Das Tierbuch in Farben. Munich and Zürich 1956

Schäfer, K.E.: Physiologische Anpassung bei Meeressäugetieren. In: Überleben auf See. II. Marinemedizinisch-Wissenschaftliches Symposium in Kiel. Kiel 1968

Scharf, K.-H. and W. Weber: Cytologie. Materialien für die Sekundarstufe II Biologie. Hannover 1976

Schlegel, H.G.: Allgemeine Mikrobiologie. 4. edn. Stuttgart 1976

Schlieper, C. In: Fortschritte der Zoologie 9, 1952

Schmeil, O. and A. Seybold: Lehrbuch der Botanik. vol. II. 57. edn. Heidelberg 1958

Schmidt, R.F. (ed.): Grundriß der Neurophysiologie. 3. edn. Heidelberger Taschenbücher 96, Berlin, Heidelberg, New York 1974

Schmidt, K.P. and R.F. Inger: Reptilien. Das Tierbuch in Farben. Munich, Zürich 1957

Schmidt, W.: Über die absolute Größe biologischer Systeme. BIUZ 14 (3), 65 – 71, 1984

Schopfer, P.: Experimente zur Pflanzenphysiologie. Freiburg 1970

Schüz, E.: Grundriß der Vogelzugkunde. 2. edn. Berlin and Hamburg 1971

Schultz, A.H.: Die Primaten. In: Die Enzyklopädie der Natur. vol. 18, Lausanne 1972

Schutzgemeinschaft Deutscher Wald: Informationen der Schutzgemeinschaft Deutscher Wald 1967

Schwerdtfeger, Fr.: Ökologie der Tiere. Hamburg 1975

Schwidetzki, I.: Die neue Rassenkunde. Stuttgart 1962

Schwidetzki, I.: Variations- und Typenkunde des Menschen. In: Bertalanffy, L.v. (ed.): Handbuch der Biologie. vol. IX. Konstanz 1965

Sengbusch, P.v.: Molekular- und Zellbiologie. Berlin 1979

Siebeck, O.: Photorezeptoren und Photorezeption. In: Bertalanffy, L.v. (ed.): Handbuch der Biologie. vol. V, Wiesbaden 1977

Siewing, R. (ed.): Evolution. UTB 748. Stuttgart 1978

Siewing, R. (ed.): Lehrbuch der Zoologie. vol. I. 3. edn. Stuttgart, New York 1980

Slijper, J.E.: Riesen und Zwerge im Tierreich. Hamburg 1967

Smolik, H.-W.: Rororo-Tierlexikon. Gütersloh 1968

Snell, F.M., Shulmann, S., Spencer, R.P. and C. Moos: Biophysikalische Grundlagen von Struktur und Funktion. vol. I. Stuttgart 1968

Souci, S.W., Fachmann, W. and H. Kraut: Die Zusammensetzung der Lebensmittel. Nährwerttabellen. Stuttgart 1981

Spector, W.S. (ed.): Handbook of Biological Data. Philadelphia and London 1956

Starck, D.: Die Evolution der Organismen. 3. edn. Stuttgart 1974

Starck, D.: Embryologie. 3. edn. Stuttgart 1975

Starck, D.: Vergleichende Anatomie der Wirbeltiere auf evolutionsbiologischer Grundlage. Berlin, Heidelberg, New York 1982

Starck, D., Fiedler, K., Harth, P. and J. Richter: Biologie. Weinheim 1981

Steitz, E.: Die Evolution des Menschen. Taschentext 16. 2. edn. Weinheim 1979

Steward, F.C. (ed.): Plant Physiology. vols. I – IV. New York and London 1959f

Stiles, W. and E.C. Cocking: An Introduction to the Principles of Plant Physiology. 3. edn. London 1969

Storch, V. and U. Welsch: Systematische Zoologie. 4. edn., Stuttgart, New York 1991

Strasburger, E.: Lehrbuch der Botanik. 32. edn. Stuttgart 1983

Stresemann, E.: Exkursionsfauna von Deutschland, vol. Wirbeltiere. Berlin 1961

Sturkie, P.D. (ed.): Avian Physiology. 3. edn. New York, Heidelberg, Berlin 1976

Tanner, W.: Altern und Tod aus der Sicht der Biologie. BIUZ vol. 10, 1980

Thimann, K.V.: Das Leben der Bakterien. Jena 1964

Ulbrich, E.: Biologie der Früchte und Samen. Berlin 1928

Vangerow, E.F.: Wandel der Erdatmosphäre und die Entwicklung des Lebens. In: Ditfurth, H.v. (ed.): Evolution. Hamburg 1975

Velhagen, K. (ed.): Der Augenarzt. vol. II. Leipzig 1972

Verbraucherdienst informiert: Grundlagen der Ernährung. Bonn 1979

Walter, H.: Einführung in die Phytologie. vol. I. Stuttgart 1963

Weber, H.: Grundriß der Insektenkunde. Stuttgart 1966

Weide, H. and H. Aurich: Allgemeine Mikrobiologie. Stuttgart 1979

Weismann, E.: Die Entwicklung und Kindheit der Tiere. Dynamische Biologie 4. Ravensburg 1976

Whittow, G.C. (ed.): Comparative Physiology of Themoregulation. vol. I. New York and London 1970

Wood, G.L.: The Guinness Book of Animal Facts and Feats. 3. edn. Enfield 1982

Wundelich, K. and W. Gloede: Natur als Konstrukteur. Leipzig 1977

Ziegler, H.: Flüssigkeitsströme in Pflanzen. In: Hoppe et. al.: Biophysik. Heidelberg 1977

Ziswiler, V.: Wirbeltiere. vols. I and II. Stuttgart 1976

Subject Index

A

Absorption maxima 123, 192
Acidity level 182
Adductor muscle 31
Adenine 195, 257
Adipose tissue 206
Age 4 ff., 133
AIDS 200
Air requirements 224
Albumen 103
Albumin 220
Aldicarb 28
Aldrin 28
Amino acid 25, 228, 235, 250, 258
Aminopeptidase 49
Ammonia 62
Amoebic dysentery 200
Anaphase 253
Anthrax 200
Aorta 216
Appendix 48 f.
Arginine 25
Arteriole 216
Artery 216
Ascorbic acid 251
Aspartic acid 25
Auditory acuity 125

B

Ballistic fruits 174
Basal metabolic rate 54, 224
Basilar membrane 243
Batrachotoxin 26
Bee venom 26
β-carotene 192
Bile production 53
Bile 211
Bird migration 38 f.
Birth weight 108 f.
Blood cells 73
Blood cells 75, 221
Blood flow 218 f.
Blood groups 222

Blood plasma 220, 222
Blood pressure 71, 217
Blood sugar level 79
Blood vessel 217
Blood vessel system 216
Blood 72 ff., 80, 206, 211, 215 ff., 219 f.
Blood volume 72
Blossom diameter 136 f.
Blossoming data 162–167
Blossoming period 161
Blossoms 158 ff.
Body hair 212
Body size 8, 10, 12–14, 17 f., 20
Body temperature 55–57, 210
Bone marrow 205
Bones 24, 208
Botany 129 ff.
Brain 205, 239 ff.
Brain surface area 115
Brain volumes 239
Brain weight 114, 240
Breast muscle 30
Breathing rate 85 f.
Bufogenin 26
Bufotenin 26

C

Calf muscle 30
Caloric equivalent 42
Cantharidin 26
Capillaries 216
Capsomer 195
Capsule 171
Carbohydrases 49
Carbohydrates 42, 227 f., 230–232
Carbon exchange 189
Cardiac cycle 215
Cardiac muscle 32
Cardinal temperature points 154
Carpal (wrist) bones 208
Catalase 250
Catapult distance 174
Cathepsin 49

Cell division 94, 253
Cell membrane 116, 252
Cell nucleus 253, 255 f.
Cell sap 180
Cell size 23
Cell 204, 251–253
Cellulose 250
Cerebral cortex 24
Cerebralization index 115
Cerebrospinal fluid 211
Cervical vertebrae 22
Chemical elements 249
Chemically defined diet 228
Chicken egg 103
Chicken pox 200
Chlordane 28
Chlorophyll 192, 250
Chloroplasts 142, 252, 256
Cholera 200
Cholesterol level 235
Chromosome sets 91, 145
Chromosomes, number of 91, 145 f.
Chymotrypsin 49
Ciliary body 241
Ciliary process 242
Circulation 65 ff., 215 ff.
Closed seasons (on hunting) 112
Clutch size 99
Coagulation time 74
Coelom fluid 80
Coenzyme I 251
Cold spots 245
Cold 241
Compound eye 122
Conduction velocity 117, 191
Cones 241
Congestin 25
Constituents 175 ff., 178
Contraction time 31
Cornea 241 f.
Coxal muscle 31
Crawling 36
Creatine 62
Creatinine 62
Cycle (of ovulation) 106
Cystine/cystein 251
Cytochrome 251, 258
Cytosine 257

D
Dart poison 26
DDT 28
DDVP 28
Dental formula 43
Development 89 ff., 94
Diaphragm 30, 32
Diastole 217
Diazinon 28
Dictyosomes 252
Dieldrin 28
Digestion 41 ff., 226 ff.
Digestive capacity 227
Digestive enzymes 49
Digestive system 236
Diphtheria 200
Diving depth 88
Diving time 88
DLM 25–27
DNA 204, 255, 257
Double membrane 252
Dryness limits 157
Dysentery (shigellosis) 200

E
Ear 118, 241, 243
Eardrum surface area 124
Eardrum 243
Effector cells 75
Egg cell 90
Egg size 96, 100 f.
Egg weight 100 f.
Egg 94
Eggs, number of 96 f.
Electric organs 117
Embryogenesis 104
Embryonal development 104
Endrin 28
Energy content 179, 227, 230–232
Energy intake 226 f.
Energy requirements 224, 229
Enteropeptidase 49
Erythrocytes 24, 75, 219 f.
Estrus 106
Evaporation 184
Excrement 62
Excretion 60 ff., 226 ff.
Eye 118–120, 122, 241 f.

Eyeball 118
Eyebrow 212
Eyelashes 212

F
Fat content 176 f.
Fat 42, 227, 230–232
Feces 211
Fenestra vestibuli (oval window) 243
Fibrinogen 250
Finger 208
Fingernail 214
Flavone 192
Flavoprotein 251
Flicker fusion rate 122
Flight distances 39
Flight speeds 38
Floral formulas 159 f.
Flying 37
Focal distance 119
Food intake 42
Food requirements 42
Foodstuffs 230
Forearm 208
Freezing point depression 63, 81 f.
Fruit(s) 138, 158 ff., 177 f.
Fruiting body formation 169
Furrow formation 253

G
Gastric juice 211
Gastricin 49
Generation time 196
German measles (rubella) 200
Germination speed 148
Germination temperature 147
Germination time 147
Germination 147 ff.
Gestation time 106
Globulin 220
Glomerular ultrafiltrate 237 f.
Glomeruli 60
Glucose 250
Glutamic acid 25
Gluthathione 251
Glycine 25, 250
Glycogen 250
Gonorrhea 200

Granulocyte 75
Gravidity 106–108
Growth speed 149 f.
Growth 147 ff., 155, 206
Guanine 257

H
Hair 212–214
Head hair 212, 213
Hearing ability 124, 243
Heart minute volume 70, 215
Heart rate 68 f.
Heart weight 65–67
Heart 65 ff., 205, 215 ff.
Heartbeat 215
Heat generation 225
Heat 241
Hectoliter weights 172
Height (of plants) 152
Hematocrit values 76
Hemin 251
Hemocyanin 250
Hemoglobin 77, 250 f.
Hemoglobin level 76
Hemoglobin quantity 221
Hemolymph 77, 80 f.
Hepatitis 200
Heptachlor 28
Herpes zoster (shingles) 200
Hibernation 59, 69, 87
Histidine 25
Homobatrachotoxin 26
Hornet venom 26
Horns 21
Hot spots 245
Human biology 201 ff.
Hydraulic conductance 186
Hydraulic conductivity 186
Hypsotoxin 25

I
Imago 94
Inactivation temperature 198
Incubation times 200
Incubation
– length of 101 f.
– dates of 102
Infectious disease 200

Influenza 200
Insect colonies 11
Insecticide 28
Insulin 250
Intercostal muscles 30
Intestinal length 47
Intestinal passage 52
Intestinal tract 48, 50 f.
Intestine 46, 206
Ions 116
Iris 241
Isoleucine 25
Jumping distances 38

K
Kidney 24, 60, 205, 237
Krause's end bulbs 212

L
Lactate/pyruvate 251
Lactation periods 111
Large intestine 48 f.
Larva 94
LD_{50} 25–27
Leaf surface area 136
Leaves 135, 138
Legumes 175
Lentil 119, 241 f.
Leprosy 200
Lethal temperature 197 f.
Leucine 25, 250
Leukocyte number 221
Lifespan 135, 173, 204
Lindane 28
Linnaeus' flower-clock 158
Lipase 49
Lipid acids 228, 235
Liver 24, 205
Liver weight 52
Load-bearing performance 32
Locomotion 29 ff.
Long-day plant 161
Lophotoxin 25
Lower leg 208
Lung weight 84
Lung 85, 205, 223
Lymph 206, 211, 222
Lymphocytes 24, 75

Lysine 25

M
Malaria 200
Maltase 49
Measles 200
Meißner's tactile corpuscles 212
Membrane, simple 252
Memory cells 75
Merkel's disk 212
Metabolism 54 ff.
Metacarpal bones 208
Metaphase 253
Metatarsal bones 208
Methionine 25
Methylene blue 251
Mevinphos 28
Microbiology 193 ff.
Mid-intestine 51
Migratory times 40
Milk 110
Milk production 111
Minerals 229
Minimum separability angle 120
Mitochondria 252, 254–256
Molecular weight 77, 250
Movement 29 ff.
Mumps 200
Murex 26
Muscle fiber 30, 116
Muscle 30–32
Muscular strength 31
Musculature 24, 205
Myoglobin 250
Myosin 250

N
Near point 242
Nectar 179
Needles 135
Nerve(s) 117, 239 ff.
Nerve fiber 116, 120
Nervous system 113 ff., 239
Nesting time 101 f.
Normal growth 108, 207
Nose 126
Nutrition 41 ff., 226 ff.
O

Odor threshold 246
Olfactory epithelium 126
Ommatidia 122
Organ size 209
Organ 210
Organelle 251 f.
Osmolarity 81
Osmotic pressure 63, 81, 180–182
Oxygen capacity 77 f.
Oxygen consumption 54, 87 f., 187, 225

P
Pain 241
Palytoxin 25
Pancreas 206
Pancreatic α-amylase 49
Parathion 28
Pelvis 208
Pepsin 49, 250
Perthane 28
Pertussis (whooping cough) 200
pH ranges 197
pH value 50 f., 154, 211
Phenylalanine 25
Photoreceptors 119 f.
Photosynthesis 188
Photosynthesis production 188 f.
Phycocyanin 192
Phycoerythrin 192
Physiological salt solutions 82
Physiological values 180 ff.
Pigment 192
Plant fibers 139
Plasma flow 180
Plasma 220
Platelets 74 f.
Poisonous effect 28
Poisons 25 f.
Polar limits 156
Poliomyelitis 200
Pollen grains 143
Prealbumin 220
Preferred temperatures 57 f.
Pregnancy 206
Pressure 241
Progeny, number of 97
Prophase 253
Protein 25, 42, 220, 227, 230–232

Protein intake 226 f.
Protein level 80
Protochlorophyll 192
Ptyalin 49
Pufferfish poison 26
Pumiliotoxin 26
Pupa 94
Purkinje cells 23 f.
Pyocyanin 251
Pyramidal cells 23

R
Rabies 200
Raw protein content 178
Reabsorption 238
Receptors 241
Rectal temperature 55
Redox potentials 251
Refractive power 242
REM sleep 240
Renal corpuscle 237
Reproduction 89 ff., 94, 158 ff.
Respiration 83 ff., 223 ff.
Respiration rate 187, 199
Respiratory quotient (RQ) 42
Resting potential 116, 191
Retina 241
Retinal 1 123
Retinal 2 123
Rhodopsin 123
Rib cage 208
Riboflavin 251
Ribosomes 252
RNA 204
Rods 241
Root depth 135
Roots 138
Rotation 190
Rotation periods 134
Ruffini's corpuscles 212
Runners (plant) 153
Running 36

S
Salamandrin 26
Saliva 50, 211
Salt concentration 64
Scarlet fever 200

Sclera 241
Sebaceous gland 212
Seeds 138, 147–149, 158 ff., 173 f., 177 f., 249
– number of 170 f.
– weight 171 f.
Senses 118 ff., 239 ff., 241
Serum albumin 250
Serum globulin 250
Sexual maturation time 105
Sexual maturity 93
Shell 103
Shoot 149 f.
Short-day plant 161
Shoulder girdle 208
Single-breath air volume 85 f.
Size 4 ff.
Skeletal muscle 215
Skeletal weight 23
Skeleton 205
Skin adnexa 212 ff.
Skin 24, 205, 212 ff.
Skull 208
Sleep requirement 240
Slenderness ratio (of plants) 152
Small intestine 48 f.
Smallpox 200
Smell 241
Snail 243
Snake venom 27
Soil 155
Somites 206
Sound level 243
Spatial discrimination threshold 245
Spawning data 96
Species, number of 5, 132
Specific weight 137 f., 209
Speed of movement 252
Speed 36 f.
Sperm 91, 211, 251
Sphincter 32
Spinal ganglion 23 f.
Spine 208
Spleen 205
Spore maturation 168
Spores 144
Sprout 138
Standard food requirements 226

Starch grains 143
Starch 188
Stimulus 118
Stomach 23, 48 f.
Stomata 140 f.
Stonefish poison 26
Stroke volume 215
Subgenual organ 118
Sugar content 179
Sugar 188
Surface load 33–35
Sweat gland 212, 214
Sweat 211
Swimming 36
Synovial fluid 211
Syphilis 200
Systole 217

T
Tarsal bones 208
Taste buds 126
Taste threshold 246
Taste 241
Teeth 21, 205, 237
Telophase 253
Temperature tolerances 182 f.
Tetanus 200
Tetrodotoxin 26
Thalassin 25
Thigh 208
Thousand seed weights 172
Threonine 25
Thymine 257
Tissue 206
Toenail 214
Toes 208
Torpor 59
Toxaphene 28
Trace elements 229
Tracheids 140
Transpiration coefficient 185
Transpiration level 184
Transpiration 185
Trees
– height 150,
– vertical growth 150
Trunk circumferences 152
Trunk diameters 151

Trypsin 49
Tryptophan 25
Tubers 178
Twitch times 30
Typhus 200

U
Upper arm 208
Urea 62 f.
Urease 250
Uric acid 62
Urine 62 f., 211, 237 f.
Urine concentration 61
Urine production 61
Urine volume 238
Uterus 31

V
Valine 25
Vater-Pacini corpuscles 212
Vein 216
Venole 216
Vernalization temperatures 148
Vertebrae, number of 22
Vessel
– diameter 139
– elements 139
Vibration threshold 127
Virion 195

Visual
– acuity angle 120 f.
– field 121, 242
– pigments 123
Vital capacity 223
Vitamin 228
– content 233
– requirements 232
Vitreous humor 241

W
Water 231
– consumption 185
– content 24, 138, 209 f., 249 f.
Weight 4 ff., 12, 14–17, 20 f.
Wing surface area 33–35
Wingbeats 32
Wingspan 15 f.
Winter rest 59
Withers, height at 19
Wood 137–139, 186

X
Xanthophyll 192
Yellow fever 200
Yolk 103

Z
Zectran 28

Name Index

A

Acacia 180
Acorn worm 9
Adder's-tongue 146, 168
Adelie penguin 78
Adenoviruses 195, 198
Admiral 95
Aepyornis (Madagascar's elephant
  bird) 20, 100
Aerobacter 197
Aesculapian snake 14, 98
Affenpinscher 18
African buffalo 21
African clawed frog 30, 256
African pancake tortoise 98
Agile frog 13, 38, 98
Agrobacterium 197
Airedale terrier 18
Albatross 15, 35, 37, 55, 99
Alder 139, 142 f., 145, 162, 171, 183, 186
Alder buckthorn 162
Alfalfa 140, 148, 154, 172, 176 f., 179, 181,
  185
Alga(e) 132, 142, 145, 181 f., 186, 187, 249
Alkali grass 165
Alligator snapping turtle 14
Alligator 54, 62, 66, 76, 79 f., 85, 88, 90, 92,
  120, 123
Almiqui (Cuban banana frog) 13
Almond 146, 171, 176, 230
Alpine ibex 21
Alpine newt 13, 254
Alpine rose 162, 185
Alpine salamander 97
Alpine squill 164
Alsimatidae 133
Amanita pantherina 144, 169
Amber snail 68
American alligator 7, 14, 58, 98
American copperhead 27
Amethyst hummingbird 34
Amoeba 54, 94, 252

Amphibian(s) 5, 6, 13, 20, 22, 26, 66, 92,
  97 f., 254
Amphiuma ("Congo eel") 13
Amur red-footed falcon 39
Anaconda 7, 14, 41, 97
Andalusian horse 19
Anemone 153, 168
Angelica 167
Angel's-tears 137, 161
Angiosperms 132
Anglo-Arabian 19
Anopheles mosquito 68
Ant 11
Antelope 36
Anthericum 158, 164
Ape 44, 115
Aphid 94
Appenzeller mountain dog 18
Apple 133, 136, 138, 140, 145, 162, 171,
  179 f., 181 f., 186 f., 230, 233
Apricot 162
Aquatic sowbug 68
Arabian 19
Arachnids 5
Arapaima 12
Archaeophis 22
Archaeopteryx 20
Archelon 20
Arctic tern 39
Arecidae 133
Arenaviruses 195
Argentavis 20
Arnica 165
Arrow worm 9
Arrowhead 164, 180
Arthropods 5, 9, 26, 82, 122
Arvicolinae 45
Asarabacca (European wild ginger) 142,
  166, 180, 185
Ascomycota (sac fungi) 132, 145, 258
Ash 134, 138 f., 143, 145, 150 f., 162, 172,
  180, 183, 186
Asp viper 27, 92

Asparagus 138, 146 f., 150, 155, 164, 230, 233
Aspergillus 198 f.
Assassin bugs 127
Aster (family) 160
Asterales 133
Asteridae 133
Atlantic white-side dolphin 115
Aurochs 21
Australian mountain ash (*Eucalyptus regnans*) 151
Australopithecus 239
Autumn crocus 146, 149, 164, 188
Avens 153
Avian flu virus 198
Avocado 172, 176
Axolotl 6, 13, 66, 92, 124
Azotobacter 197, 199

B
Baboon 7, 16, 60, 92 f., 105, 108, 110, 115
Bacillus 195–199, 252, 257
Bacteria 132, 195–199, 257
Bacteriophage λ 195, 256 f.
Bactrian camel 7
Baker's yeast 230, 258
Bald cypress 152
Baluchitherium 21
Bamboo 149, 152
Banana 136, 145, 149, 156, 184, 230, 233
Baobab 152
Barbel 96
Bare-toothed russula 145
Barley 135, 138, 143, 145, 147 f., 154, 156 f., 165, 172 f., 175, 177 f., 180, 183–185, 188, 255
Barn owl 7, 15, 35, 99, 121
Barn swallow 35, 38–40, 99 f., 102 f.
Barracuda 12
Barzoi 18
Basidiomycota 132, 144
Basket willow 163, 186
Basking shark 12
Bat 30, 37, 43, 46 f., 52, 54, 56, 59 f., 69, 72, 76, 85, 92 f., 107 f., 115, 118, 120, 124, 126
Bauhinia purpurea 174
Bay boletus 144, 169

Bay laurel 135, 176
Beadlet anemone 6
Beaked sea snake 27
Beaked whale 88
Bean 136, 140, 148, 154, 172 f., 175 f., 184 f., 187, 230
Bear 41, 43
Bearberry 185
Bearded darnel 165
Bear's breeches 174
Beaver 7, 16, 43, 88, 92, 106
Bedbug 6, 51, 57
Bedstraw 146
Bee orchid 145
Bee 6, 30, 32, 37, 51, 90 f., 120, 122, 127
Beech (family) 159
Beech marten 45, 93, 107
Beech 136, 143, 150, 154, 173, 184 f.
see also European beech
Beechnut 176
Beefsteak or brain mushroom 169
Bees and ants (hymenoptera) 5, 10
Beet (*Beta vulgaris*) 142, 177, 188
Beet 148
Beetle 5, 10, 122
Beetroot 156
Belladonna 167
Bellflower 190
Beluga whale 84
Beluga 12
Bent grass 162, 165
Berberis 140
Bilberry 155
Birch (family) 159
Birch 133 f., 137, 139, 143, 145, 151, 159, 162, 170 f., 180, 185 f.
Bird cherry 163
Birds 5, 7, 15, 20, 22 f., 34, 50, 55, 66, 84, 92, 99 f., 102, 104, 115, 119, 127
Bird's-nest orchid 164
Bishop's goutweed 141, 153, 166
Bison 7, 16, 36, 67, 93
Bitterling 96 f.
Black alder 134, 137, 151
Black bear 110
Black grouse 99
Black medick 166
Black mold 257

Black morel 145, 169
Black poplar 134, 144, 151, 163, 170
Black redstart 40, 79, 100–102
Black rhinoceros 16, 21
Black wildebeest 21
Black woodpecker 15, 99
Blackberry 162, 179, 182, 233
Blackbird 7, 34 f., 55, 66, 92, 99–102, 120
Blackcap 40, 99, 103
Black-eyed pea 177
Black-headed gull 55, 99–101, 103
Blackthorn 163
Bladder wrack 183, 187
Bladderpod 158
Bloodhound 18
Blowfly 32, 37, 54, 94, 122, 127, 254
Blue mussel 54, 68, 77, 80 f., 232, 234
Blue shark 12, 97
Blue tit 15, 41, 99, 102
Blue whale 16, 46 f., 62, 67, 72, 88, 93, 106,
    108, 110, 114, 119 f.
Bluebell (family) 159
Blue-headed wagtail 38
Blusher 169
Boa constrictor 14, 66
Boa(s) 7, 22
Body louse 57
Bog rosemary 163
Bonito 76
Bony fish 63, 80
Borage (family) 160
Borage 179
Bordetella 195
Bowhead whale 16, 88, 119
Boxer 18
Boxwood (common box) 138, 162
Brachiosaurus 20
Brackenfern 145, 168, 183
Brazil nut 172
Brazilian porcupine 108
Bread mold 254
Bream (Carp) 96
Brimstone 95
Brine shrimp 64
Bristlecone pine 134
Bristleworm 64, 119
Bristly oxtongue 158
British yellowhead 167

Brittle stars 9
Broad bean 135, 147, 150, 172, 175, 178, 253
Brome 162, 165
Bromeliales 133
Brontosaurus 20
Brook lamprey 96
Broomrape 171
Brown algae 132
Brown bear 7, 46 f., 61, 93, 105 f., 111
Brown bullhead 125
Brown rat 16, 69, 108 f., 119 (see also Rat)
Brown roll-rim 169
Brown trout 96 f.
Brussel sprouts 230
Bryony 150, 168, 190
Bryophyllum 162
Buchfink 7, 35, 39, 54, 66, 73, 76, 99–102,
    120, 124
Buckbean 166
Buckwheat 141
Budgerigar 67, 84, 99, 101 f., 125
Buffalo 110
Bugle 166
Bull thistle 166
Bullfrog 38, 66, 124, 258
Bullhead 96
Bumblebees 11, 30, 32 f., 37, 80
Burbot 96
Burdock 146, 166
Burnet 167
Bur-reed 164
Bush vetch 146
Bushmaster 14, 27
Buttercup (family) 159
Buttercup 136, 153, 161, 166, 173
Butterfly bush 170
Butterfly 5, 10, 54, 95
Button mushroom 145, 230, 234
Buzzard 7, 35, 41, 85
see also Common buzzard
Byrum 183

C
Cabbage 138, 140, 149, 173, 176, 230, 233
Cacao 156 f., 176
Cactus 137, 141
Caddis fly 32
Caecilian 13, 20

Caiman 69
Calliphora 33
Camargue horse 19
Cambarus 116
Camel 16, 36, 43, 46 f., 52, 54, 57, 61, 93, 105, 107, 110
Campanula 135 f., 141, 145, 166, 170
Campanulales 133
Campion 161, 167
Canada thistle 170
Canadian waterweed 141 f., 146, 164, 180 f., 183, 188
Canary 7, 71, 79, 92, 101, 124
Candida 198 f.
Caoutchouc 156 f.
Capuchin monkey 7, 56, 67, 69, 86, 92
Caraway 146, 166
Carcharodon 20
Carcinus 116
Carline thistle 135, 167
Carp 6, 12, 54, 58, 62, 64, 66, 69, 73, 76, 78, 81, 87, 92 f., 96 f., 114, 117, 119, 123, 231, 256
Carrion beetle 127
Carrion crow 15, 22, 34, 55, 67, 73, 76, 84, 86, 99–102
Carrot (family) 159
Carrot 138, 146 f., 148 f., 167, 170, 155, 173, 180, 188, 230, 233
Cartilaginous fish 63
Caryophyllales 132
Cassava 156 f., 178
Castor seed 176
Cat shark 69, 76, 79, 81, 86, 92, 97, 126
Cat 7, 23, 30–32, 43, 46–48, 52–54, 56, 60 f., 63, 69, 71–76, 78, 81, 84–86, 91, 93, 106 f., 109–111, 114, 116–119, 121 f., 124–126, 253
Cat's-ear 158
Cattail 164, 181
Cattle 7, 23 f., 41, 45–54, 56, 60–63, 69–72, 74, 76, 78–81, 84, 86, 90–93, 106 f., 109–111, 114 f., 118 f., 124, 126, 252 f., 255–257
Cauliflower 147 f., 230
Cauliflower mushroom 169
Cave bear 21
Cedar 134, 151 f.

Celery 149, 173, 182, 230
Centaury 161, 167
Centipede 6, 9, 36
Cephalopods 5 f., 9, 87
Chameleon 58, 120
Chamois 7, 16, 43, 105, 107, 112
Chamomile 146, 166
Chanterelle 144, 169
Chara 191, 252
Charadriidae 39
Cheetah 36
Cherry laurel 135, 186
Cherry 133, 137, 140, 146, 171, 230, 233
Chicken 7, 23 f., 41, 46 f., 50, 52–55, 60, 62, 67, 69–72, 74–76, 78–80, 85, 87, 92 f., 103 f., 118–120, 123–126, 253, 256–258
Chicory 158, 167
Chiffchaff 40, 99, 101–103
Chihuahua 18
Chimpanzee 7, 16, 52, 56, 71, 73, 76, 78, 84, 86, 92 f., 105–108, 111, 114 f., 121, 125, 239
Chlamydomonas 182, 251
Christmas rose 145, 166
Chrysanthemum 162
Chub 96
Cicada 10, 125
Cinquefoil 153, 166
Citrus 156
Cladophora 187
Cleavers 166
Clematis 163
Clostridium 195–198
Clover 143, 171 f.
Clusterhead 146, 166
Clydesdale 19
Cnidaria 5, 8, 25, 87
Cobra 7, 27
Cocci 197
Cockatoo 7
Cockchafer (May bug) 32, 37, 47, 122, 127
Cocker spaniel 18
Cockle 64
Cockroach 5, 10, 31, 51, 68, 91, 94, 116–118, 123, 127
Coconut 146, 171, 173, 176, 230
Coconut palm 136, 139, 150, 156 f., 170

Cod (Baltic Sea) 74, 81, 254
Cod 12, 96 f., 234
Coelacanth 90
Coffee 156 f., 173, 176
Collie 18
Colombian giant toad 13
Colorado potato beetle 90, 250
Colossochelys 20
Coltsfoot 146, 153, 166, 170
Colubrid (snake) 76
Comfrey 165
Comma 95
Commelinidae 133
Common agrimony 161, 167
Common alkanet or bugloss 167, 170
Common aspen 133, 137, 144, 155, 163, 173
Common barberry 163
Common bean 143, 145, 166
Common bird's-foot trefoil 166
Common blue (butterfly) 33
Common bugle 153
Common butterwort 166
Common buzzard 15, 34, 37, 55, 67, 69, 84,
    99–101, 103
Common chickweed 170
Common clubmoss 168
Common earthball 144, 169
Common eider 36
Common eland 16, 21
Common Eurasian spadefoot toad 13, 92,
    98
Common evening primrose 149, 161, 167,
    173
Common frog 13, 59, 69, 72 f., 76, 92 f.,
    97 f., 120, 124
Common fumitory 166
Common glasswort 167, 181
Common hemp-nettle 166
Common hermit crab 120
Common hogweed 165, 172
Common house mouse 16, 23, 36, 54, 58,
    67, 92, 106–109, 118, 123
    see also Mouse
Common kestrel 35, 41, 67, 103, 121
Common lizard 14, 59, 98
Common madder 153
Common mare's tail 167
Common milkwort 166

Common moonwort 133, 168
Common or Harbor seal 7, 16, 23, 43, 47,
    60 f., 63, 71, 76, 85, 88, 93, 107–109,
    111
Common periwinkle 6
Common pipistrelle 69, 88
Common polypody 168, 183
Common quail 39 f.
Common raven 7, 15, 67, 99–101, 115
Common redstart 40, 102
Common reed 144, 165
Common shrew 47, 108
Common sow thistle 158, 166
Common spikerush 164
Common spotted orchid 170
Common swift 7, 15, 34 f., 37, 39 f., 55, 67,
    69, 79, 86, 99–101, 103
Common teal 39
Common toad 6, 13, 58 f., 69, 73, 85, 92,
    97 f., 256
Common vetches 172
Common vole 16, 36, 52, 58, 84, 90, 107
Common wall lizard 14, 59, 98
Common yarrow 144, 146, 153, 167
Condor 7, 15
Coral snake 27
Coral 8
Cordgrass 165
Corn (maize) 136, 138, 140 f., 143, 146–149,
    154 f., 172 f., 175–178, 181, 184 f., 188,
    255
Corn buttercup 170
Corn cockle 166, 170
Corn poppy 146, 166, 170
Corn seed 172
Corn spurry 167, 173
Cornelian cherry 163
Cornflower 146 f., 166, 170, 173
Coronaviruses 195
Corydalis 166
Corynebacterium 195
Cosmos 162
Cotton grass 142, 164
Cotton 139, 156 f., 176
Cow 258
Cowberry (lingonberry) 135, 163, 182
Cowslip 167
Cow-wheat 167

Coyote 36, 61
Crab 257
Cranberry 161
Crane fly 33
Crane 7, 38–40
Cranesbill 167, 174
Crayfish 6, 54, 57, 63, 81
Creeping Jenny 167
Cress 233
Crest (or kinglet) 15, 35, 41
Crested dog's-tail 165
Crested lark 102
Cricket 51, 124
Crimson clover 148
Crinoids (feather-stars) 9
Crocodile 14, 20, 22, 71 f., 78, 93, 121
Crocus 164
Cro-Magnon man 239
Cross spider (Araneus) 68
Crossbill 39
Crossed viper 14, 27, 58 f., 69, 92, 97 f.
Cross-leaved heath 145, 163, 183
Crow 35, 69
        see also Common raven
Crow 7, 69, 124
Crowberry 163
Crown vetch 185
Crucian carp 96
Crustacean 5 f., 9, 87, 117
Ctenophore 8
Cuban tree frog 13
Cuckoo flower 167
Cuckoo 7, 39 f., 99, 101
Cuckoopint 142, 164, 187 f.
Cucumber 136, 145, 148, 154, 166, 179, 230,
        249
Currants 233
Cuttlefish 6, 70 f., 80, 91, 116, 232
Cyclamen 133, 136, 143, 145, 161, 165
Cyclops (water flea) 91
Cyperales 133
Cypress (*Cupressus*) 134, 151
Cypress spurge 135, 141, 146, 168

D
Dachshund 18
Daffodil 164
Daisy 142, 166
Dalmatian 18
Damselfly 33
Dandelion 135, 142, 146 f., 158, 166, 170,
        173
Daylily 158, 161
Dead-nettle 167, 170, 179
Death cap 144, 169
Deer fern 168
Deer 7, 8, 36, 43–47, 60, 68, 93, 105, 107,
        110–112, 126
        see also Red deer (wapiti)
Dentaria glandulosa 153
Desert monitor 24
Desert plants 135
Devil anglerfish 79
Diatoms 132, 251 f.
Dice snake 14, 98
Dicotyledoneae 132
Dictamnus 162
Dill 161
Dilleniidae 132
Dingo 18
Dinosaur 20, 22, 36
Dinotherium 21
Diplococcus 196–198
Diplodocus 20
Diptera 10
Dock 165, 188
Dodder 167
Dodo 20
Dog 7, 18, 23, 31, 43, 46–48, 50, 52–54, 56 f.,
        60–63, 67, 69–76, 78–81, 84 f., 87, 90,
        92 f., 106 f., 109–111, 114, 116, 118,
        121, 124–126, 253, 256, 258
Dog fennel or chamomile (Anthemis) 166
Dog flea 57
Dog rose 133, 136, 163
Dog violet 166, 174
Dolichovespula wasp 11
Dolphin 7, 16, 36, 47, 60, 69, 76, 80, 85, 106,
        108, 110, 114, 124
Domestic pig 7, 93
Domestic pigeon 66
Donkey 7, 107, 109–111, 120, 258

Dormice 43
Douglas fir 134, 137, 150, 162, 171
Dragonfly 5, 10, 30, 33, 37, 68, 91, 122
Driver ants 11
Dromedary 7, 36, 56, 106
Drosophila, see Fruit fly
Dry rot fungus (Serpula lacrymans) 144
Duck 7, 50, 52 f., 55, 57, 60, 66, 69, 71 f.,
    75 f., 79, 85, 88, 92 f., 123 f., 126
Duckweed (family) 160
Dung beetle 47
Earth boring 34
Dunnock 102
Dwarf gecko 14
Dwarf goby 12
Dwarf honeysuckle 161
Dwarf lemur 16
Dyer's greenweed 135

E
Eagle 7, 35, 114
Eagle owl 7, 15, 67, 99 f., 102, 125
Earthworm 6, 31, 54, 76 f., 81 f., 91, 116 f.
Earwig 6, 10, 57, 122, 127
Eastern white pine 137, 141, 173
Echidna 56
Echinoderms 5, 9, 87
Ectoprocta (Moss animals) 9
Edelweiss 166
Edible crab 68, 77
Edible frog 73 f., 78, 81, 92, 97 f., 256
Edible snail 6, 31, 37, 68, 77, 80, 91, 93, 232,
    250
Eel 6, 12, 22, 36, 54, 69, 71–73, 76, 81, 123 f.
Eelgrass 146, 164
Egret 35, 92
Elderberry 163, 233, 251
Electric catfish 117
Electric eel 78, 117
Electric rays 52, 66, 69, 71, 73, 79, 86, 97,
    117, 119
Elephant 7, 16, 21, 23, 36, 41, 43, 46 f., 52,
    54, 56, 62 f., 67, 69, 73, 84 f., 93, 107 f.,
    110 f., 114 f., 118, 120
Elephant fish 117
Elephant seal 16, 69, 88
Elm 134, 137, 140, 144, 151, 171, 186
Emperor penguin 100

Enchanter's nightshade 166
Endive 148
English elm 152, 162
English holly 135, 141, 155, 163, 181
English oak 141, 146, 151
English or narrowleaf plantain 167
Enterobacteria 196
Enteromorpha 191
Ergot 145
Ericales 132
Ermine 16, 44, 105
Escherichia 196–199, 251, 255–257
Etruscan shrew 16, 108
Eucalyptus 152
Euglena 145, 252, 254, 257
Euphorbiales 133
Eurasian aspen 186
Eurasian badger 17, 44, 60, 105 f., 112
Eurasian black vulture (Monk vulture) 15
Eurasian bullfinch 102, 124–126
Eurasian coot 34, 99–102
Eurasian curlew 15, 40, 102
Eurasian dotterel 39
Eurasian jay 15, 66, 99, 102
Eurasian minnow 96, 120, 124
Eurasian nuthatch 99, 103
Eurasian pygmy owl 15
Eurasian reed warbler 99, 103
Eurasian siskin 39, 85
Eurasian water shrew 46 f., 115
Eurasian woodcock 15, 112
European beach or marram grass 165, 181
European beech 134, 138, 140, 146, 151,
    155, 159, 163, 172, 183, 185 f.
European black pine 137
European columbine 165
European conger 12, 86
European fire-bellied toad 98
European goldfinch 103
European greenfinch 102
European ground squirrel (suslik) 61, 88
European hare 7
    see also Hare
European hepatica 136, 140, 161, 166
European hornbeam 133 f., 138 f., 143,
    150 f., 159, 163, 171, 180, 186
European kingfisher 37, 99–101
European nightjar 103

European perch 12, 96
European pond turtle 7, 30, 47, 92 f., 98, 121
European rabbit 68
European rhinoceros beetle 51
European rowan 133, 137, 146, 162
European serin 102
European smelt 96
European speedwell 165, 185
European spindle 146, 163, 186
European starling 7, 34 f., 37, 39, 41, 55 f., 67, 71, 99 f., 102 f., 125 f.
Exmoor pony 19
Eyed hawk moth 95
Eyed or Jeweled lizard 14

F
Fabales 133
Fagales 132
Fairy flax 167
Falcon 37
Fallow deer 7, 112
False lily of the valley 164
False medlar 163
False morel 145
Fat dormouse 7, 17, 44, 59, 72, 88, 105, 107
Feather grass 165
Fern tree 150
Fernlike plants 132
Ferns 168, 183, 190
Fescue 165
Fiddler and ghost crabs 122
Field bindweed 135, 153, 165, 170, 185
Field cricket 57, 127
Field horsetail 153, 168
Field marigold 158
Field mushroom 145, 169
Field or hedge maple 145, 150 f., 162, 183, 186
Field rose 161
Fieldfare 40
Fig 134, 180
Figwort (family) 160
Figwort 165
Fin whale 17, 36, 67, 85
Fink 31
Fir 134 f., 144, 146, 150–152, 154 f., 172, 181, 188, 255

Fire salamander 6, 13, 66, 73, 85, 92, 97 f., 256
  see also Salamander
Fire-bellied toads 97
Firebug 91
Firecrest 103
Firefly 122
Fireweed 153, 166, 179
Fish 5 f., 12, 20, 26, 64, 66, 92, 96 f.
Fjord horse 19
Flamingo 22, 99
Flatworm 5, 8
Flax 139, 161, 173, 176–178, 185, 188, 252
Flea 10, 38
Flint corn 172
Flower fly 127
Flowering rush 164
Fly agaric 144, 169
Fly 5, 258
Flyingfish 37
Fodder beets 180
Forget-me-not 144, 146 f., 167, 170
Forsythia 163
Fox 7, 17, 38, 44, 47, 56, 60 f., 67, 69, 76, 85, 93, 105–107, 110 f., 115
Fox terrier 18
Foxtail grass 143
Fragrant orchid 164, 171
Francisella 196
Freshwater (pearl) mussel 6
Freshwater turtle 125
Frog orchid 164
Frog 22, 24, 30–32, 54, 60, 62 f., 70 f., 79 f., 82, 87, 90, 114, 116 f., 119, 121–123, 250, 255
Frog's bit 164, 180
Frostweed 136, 141, 161, 167
Fruit fly 6, 68, 82, 90 f., 94, 121, 253, 255 f.
Frullania 183
Fungus 132, 169, 197–199, 257
Funnel (or Agaric) 145
Fusarium 197–199

G
Galapagos giant tortoise 7, 97
Galinsoga 170
Garden loosestrife 166
Garden tiger (butterfly) 95

Garden warbler 7, 40, 102
Garganey 40
Garlic mustard 166
Gazelle 36
Gecko 58, 122
Gemsbok 21
Gentian (family) 159
Gentianales 133
Gentoo penguin 36
Geraniales 133
Geranium (family) 160
Geranium 140
Gerbil 24
German shepherd 18
Germander 182
Gharial 14
Giant clam 6
Giant moa 20
Giant neotropical toad 13
Giant salamander 7, 13
Giant Schnauzer (Riesenschnauzer) 18
Giant squid 36, 118
Giant tortoise 36, 58, 69
Gibbon 7, 17, 38, 92 f., 107, 114 f.
Ginkgo 142, 145, 155
Giraffe 7, 17, 36, 44, 69, 71, 85, 93, 107, 110
Globe flower 167
Glycine 190
Gnu 36
Goat willow 133, 163
Goat 8, 45 f., 48, 52 f., 56, 62 f., 69, 71–76,
     78 f., 93, 108–111, 115, 126
Goldcrest 99, 101
Golden eagle 15, 99 f., 102
Golden ground beetle 6
Golden hamster (or Syrian hamster) 7, 17,
     52, 61, 69, 71, 73 f., 76, 78, 84 f., 92 f.,
     104, 106–109, 114, 120, 254
Golden oriole 40, 99, 103
Golden plover 39
Golden saxifrage 166
Goldenrod 153
Goldfish 6, 30, 54, 58, 64, 69, 78, 92, 123 f.,
     256
Goliath frog 13
Good King Henry 166
Goose 7, 23, 37, 55, 60, 66, 69, 76, 80, 256
     see also Greylag goose

Gooseberry 163
Goosefoot (family) 159
Gorilla 8, 17, 52, 85, 92 f., 107 f., 111, 114,
     124, 239
Gorse 146
Goshawk 15, 35, 37, 55, 57, 99, 115
Grain weevil 250
Grain 175
Grant's zebra 19
Grape vine 133, 135, 190
Grape 230
Grapefruit 182
Grass snake 14, 22, 36, 54, 58 f., 69, 71–74,
     98
Grasshopper warbler 102
Grasshopper 5, 10, 30, 38, 51, 94, 118, 124
Gray heron 7, 34, 40, 99 f., 102
Gray partridge 37, 99, 103, 112
Gray seal 8, 88
Gray whale 36, 258
Grayling 96
Great black-backed gull 15, 35, 38
Great bustard 15, 35
Great cormorant 7, 37, 88, 99
Great crested grebe 7, 88, 99–101
Great Dane 18
Great diving beetle 36, 47, 57, 91
Great Horse 19
Great ramshorn snail 77
Great reed warbler 40, 102
Great spotted woodpecker 15, 35, 55, 66,
     99–102
Great tit 7, 35, 67, 99–101, 103
Greater celandine 167, 170
Greater duckweed 164
Greater siren 13
Greater weever fish 86
Green algae 132
Green lacewing 33, 37
Green toad 13, 66, 92, 97 f.
Green turtle 14
Green woodpecker 102
Greenbottle blowfly 51
Greenland shark 12
Greenland wheatear 39
Green-veined white 94
Grevy's zebra 19
Greyhound 18, 36, 68

Greylag goose 35, 40, 92, 99 f.
Griffon vulture 7, 15, 35, 99–101
Grizzly bear 17
Ground beetle 57, 82
Ground cedar 168
Ground ivy 153, 166
Groundsel 170
Grouper 12
Gudgeon 96
Guenon 114 f.
Guinea pig 8, 17, 30 f., 44, 48, 52–54, 56–58,
    63, 67, 69, 71–76, 78, 84, 86, 90–93,
    106–111, 114, 122–124, 253–256
Gull 57
Guppy 6, 58, 92
Gymnosperms 132
Gypsophila 180
Gypsy moth 95
Gypsy 169

H
Haddock 231
Haemophilus 196
Haflinger 19
Halibut 12
Halophytes 182
Hamamelididae 132
Hamster 17, 44, 53, 59, 87
Harbor porpoise 61, 62, 78, 86, 88, 114 f.
Hare 36, 44, 54, 60, 67, 72, 92 f., 105, 107,
    112, 119
Hartmann's zebra 19
Hart's tongue fern 168, 183
Harvest mouse 68
Harvester ant 26
Hawk moth 37
Hawk 7, 35, 41, 85
Hawksbill turtle 92
Hawkweed 141, 158, 166
Hawthorn 163, 186
Hazel dormouse 17, 44, 56, 59, 88
Hazel 143, 159, 163, 185
Hazelnut 133, 173, 176, 186, 230
Head louse 91
Heath (family) 159
Heath 133, 146, 163
Hedge false bindweed 137, 144, 168, 190
Hedge mustard 167, 170

Hedgehog 8, 17, 44, 46 f., 56 f., 59 f., 67, 69,
    72, 78, 85, 88, 91–93, 107, 111, 114 f.,
    124, 169, 255
Heliozoa 91, 253
Hellebore 135, 146, 164, 167, 181
Helleborine 164
Heloderma 22, 26
Hemp agrimony 167
Hemp 139, 147, 149, 154, 173, 176
Henbane 161, 165, 170 f., 180
Hens and chicks 166
Herb Paris 164
Herb Robert 167, 170
Herpesviruses 195, 198, 257
Herring gull 7, 34 f., 67, 99, 115
Herring 6, 22, 93, 96 f., 231, 257
Heteroptera 10
Hibiscus 137, 161
Hickory 138
Hippopotamus 8, 17, 21, 23, 36, 44, 47, 52,
    56, 84, 88, 93, 105, 107 f., 114
Hoary alyssum 166
Hoary cress 167
Holm oak 181
Homing pigeon 37
Homo erectus 239
Homo heidelbergensis 239
Homo sapiens neanderthalensis 239
Homo sapiens 239
Honesty 167
Honey buzzard 38
Honey fungus 144, 169
Honeybee 11, 33, 94, 122 f.
    see also Bee
Honeysuckle (family) 159
Hooded crow 38
Hoopoe 40, 103
Hopping mouse 61
Hops 143, 166, 190
Horn of plenty 169
Hornets 11, 37
Horse chestnut 137, 139, 144, 146, 155, 163,
    173, 176, 179, 186
Horse mushroom 169
Horse 8, 17, 19, 23, 36, 44–54, 56, 61, 63,
    68–76, 78–81, 84–87, 90, 92 f., 106 f.,
    109–111, 114 f., 118, 121, 124, 257 f.
Horsefly 33, 37

Horseshoe bat 7, 43
Horseshoe crab 123
Horsetail 146
Horsetails (family) 132
Horseweed 170
Hound's tongue 166, 179
Houndshark 52, 79
House cat 36, 92
House cricket 82, 90, 94, 255
House martin 103
House sparrow 7, 34, 37, 66, 84, 92, 99–101
House spider 91
Housefly 6, 30, 33, 37, 91, 94, 122 f., 254
Human 8, 22–24, 30, 32, 36, 38, 44, 46 f., 50,
    52–54, 56 f., 60 f., 63, 67, 69–76, 78–81,
    84–88, 90–93, 105–110, 114 f.,
    117–127, 251 f., 254–258
Hummingbird hawk-moth 33, 37, 95
Hummingbird 7, 31, 34, 41, 54, 67, 76, 84,
    86, 99–101
Hydra 91
Hydrodyction 191
Hydrophytes 181
Hydrozoa 8
Hyena 44, 61, 63
Hygrophorus marzuolus 169
Hypnum 187

I
Ibex 8, 17, 44, 47
Iceland moss 187
Icelandic poppy 158
Ichthyosaurs 20
Ichthyosaurus 20
Iguanodon 20
Impala 21, 38
Imperial eagle 15
Indian pony 19
Indian rhinoceros 17, 21
Influenza virus 198, 251
Ink cap 150
Inocybe 144, 169
Insects 5, 10 f., 32 f., 82, 94
Invertebrates 6, 54, 57, 64, 68, 77, 91
Iridoviruses 195
Iris 137, 140, 253
Ironwood 138
Isopod 82

Italian greyhound 18
Ivy (family) 159
Ivy 133, 135, 143, 152, 162, 179 f., 185 f.
Ixodes ricinus tick 91

J
Jackdaw 39, 99–102
Jaguar 67
Jasmine 179, 190
Jellyfish 80
Jerboa 38
Juniper 133, 135, 146, 151, 163
Jute 139

K
Kalanchoe 162
Kangaroo rat 61
Kangaroo 8, 17, 37, 38, 54, 56, 71, 92, 107 f.,
    110, 258
Katydid 10, 58, 124
Kernel corn 138, 156 f.
Kiang 19
Kidney vetch 168
King cobra 14
King penguin 55, 99, 101
Kinglet (or crest) 15, 35, 41
Kitefin shark 52
Knotweed (family) 159
Knotweed 136, 167
Koala 17
Kohlrabi 149
Komodo dragon 14
Krait 27
Kudu 21
Kulan 19

L
Laburnum 163
Lactobacillus 196, 198 f.
Lady fern 168
Ladybug 33
Lady's mantle 166
Lady's slipper 145, 164
Lama 8, 44, 73
Lamprey 73 f., 77 f., 80, 90, 92, 123
Lancelet 6, 90 f.
Larch boletus 144, 169
Larch 134, 137, 146, 151, 155, 163, 181, 186

Large pink 167
Large yellow underwing 95
Large-leaf linden 137, 151, 163
Lark 7
Larkspur 147
Latimeria 12
Lavender 179
Leach 91
Leafcutter ant 11
Least weasel 17, 38, 41, 44
Leatherback turtle 14, 88, 97
Leech 6, 9, 249
Leek 188, 230
Leguan 14, 47
Lemon balm 153
Lemon 136, 173, 181 f., 233
Lentil 143, 147, 149, 154, 172, 175
Leopard 17
Leopard's bane 174
Leptospires 196
Leptotricha 196
Leptotyphlops 15
Lesser butterfly orchid 164
Lesser celandine 167, 255
Lesser Egyptian jerboa 24
Lesser rattlesnake plantain 164, 171
Lesser whitethroat 39 f., 102
Lettuce 138, 147, 149, 173, 230, 233
Leuconostoc 199
Liana 139, 186
Lichens 132, 183, 187
Lilac 136, 140, 143, 145, 163, 181
Liliales 133
Liliidae 133
Lily (family) 160
Lily 140, 188
Lily of the valley 136, 142, 164
Linden (family) 160
Linnet 102
Lion 17, 23, 36, 38, 41, 44, 46 f., 52, 67, 69,
    84, 86, 93, 107, 111, 114, 118 f.
Lippizaner 19
Liriodendron (tulip tree) 136
Little egret 115
Little owl 41, 67
Little woodstar 15
Little-leaf linden 137, 146, 163
Livelong saxifrage 167

Liver fluke 91
Livid pinkgill (*Entoloma sinuatum*) 144,
    169
Lizard 7, 14, 54, 72, 80, 93, 121, 124
Lizard-toed gecko 14
Loach 96
Lobaria 187
Lobster 6, 31, 54, 63, 68, 71, 77, 80 f., 93,
    117, 123, 232
Loligo 116 f.
    see also Squid
Long beech fern 168
Long-beaked echidna 8, 56 f., 76
Long-eared bat 87
Long-eared owl 100, 102 f., 125
Loon 88, 115
Loose silky-bent 165
Lotus 137
Louse 51, 94
Lugworm 54, 68, 77, 80
Lung lichen („lungwort") 183
Lungfish 6, 12, 78, 85 f., 90
Lungwort 166, 185
Lupine 135, 147, 149, 166, 172, 174, 178
Lycopodiophyta 132, 133, 145, 153, 168
Lynx 44

M
Macaque 24, 92 f., 107
Mackerel 6, 78, 96
Magnolia 137
Magnoliidae 132
Magpie 7, 34 f., 99–102
Mahogany 137
Maidenhair spleenwort 168
Makoré 137
Male fern 142, 144, 146, 168
Mallard 34 f., 37, 67, 84, 99 f., 102 f., 112,
    115
Mallow (family) 160
Malvales 132
Malve 166
Mamba 15, 27, 36
Mammals 5, 7, 16, 21, 43, 48–50, 56, 61 f.,
    67, 84, 87, 104, 105 f., 109–111, 114 f.,
    256
Mammoth 21
Manatees 22

Man-eating shark 12
Mangrove 177, 181
Manila hemp 139
Manta ray 12, 97
Map butterfly 95
Maple (family) 159
Maple 139, 142 f., 154, 172, 186
Marabou stork 15
Marchantia 252
Marine iguana 14, 58
Marlin 12, 36
Marmoset 105, 108
Marmot 17, 44, 56, 59, 88, 105
Marsh cinquefoil 167
Marsh clubmoss 168
Marsh grass of Parnassus 161
Marsh horsetail 168
Marsh marigold 136, 141 f., 146, 167, 185
Marsh or Lake frog 13, 66, 124
Marsh tit 103
Marsh warbler 103
Martagon lily 164
Marten 112
Masterwort 167
Mastodonsaurus 20
Mat grass 165
Mayfly 10, 33 f.
Meadow ant 11, 122
Meadow cranesbill 161
Meadow grass 255
Meadow horsetail 168
Meadow pipit 39
Meadow sage 146 f., 167
Meadowsweet 166, 170
Meal moth 51, 82, 94
Mealworm 54
Megaloptera 10, 33
Melick 165
Melon 143, 154 f., 173
Mercurialis 140, 153, 165
Mergus 37
Metabacterium 196
Mezereon 163, 186
Micrococcus 196, 199, 251
Midge 33, 77
Midwife toad 13, 92, 97 f.
Migratory locust 33, 37, 68, 70, 91, 254, 257
Military orchid 140, 164

Millet grass 165
Millet 156, 173, 175, 178, 184
Mimosa 191
Miniature poodle 18
Minke whale 47
Mint (family) 160
Mistle thrush 103
Mistletoe 140, 146, 163, 181
Mitten crab 63 f.
Mold 150, 180
Mole 8, 17, 36, 41, 44, 46 f., 56, 60, 67, 92 f.,
    107
Mollusks 5, 26
Monilia 198 f.
Monitor lizards 22
Monkey 45, 123
Monkshood 166
Monocotyledoneae 133
Moon jelly 31, 64, 90 f., 117, 249
Moor frog 13, 98
Moorhen 103
Moose 8, 17, 105
Moray eel 86
Moschatel 167
Mosquito 33, 37, 82, 91, 94
Moss 132 f., 142, 183, 187
Mountain arnica 141
Mountain avens 133
Mountain pine 135, 137
Mourning cloak 95
Mouse 8, 24, 30, 41, 44, 46 f., 52 f., 56, 60,
    62 f., 69, 71–76, 78, 85–87, 91, 93, 109
    f., 114, 124, 253, 255 f.
Mouse-ear chickweed 166, 180
Mucor mold 150, 180
Mucor 197 f.
Mudpuppy 47, 72 f.
Mugo pine 163
Mugwort 170
Mule 8
Mumps virus 199
Musk deer 73
Muskrat 17, 48, 88, 115
Mussel 5, 9, 77, 87
Mustard (family) 160
Mustard 146, 149, 167, 173, 179
Mute swan 15, 22, 34 f., 99–102
Mycobacterium 196–199, 257

Mycoplasmas 196
Myotis (bat) 8, 43, 67, 88

N

Narrowleaf bittercress 174
Nasturtium 142, 147, 179
Natterjack toad 13, 73, 92, 98, 124
Neisseria 196–199
Nettle (family) 159
Nettle 143, 153, 165, 180, 252
Neuroptera 10
Neurospora 199
New forest pony 19
Newfoundland 18
Newt 60, 98, 253
Nightingale 40, 99–101, 103
Nightshade (family) 160
Nightshade 140, 170
Nile crocodile 14, 47, 98
Nilwaran 98
Nitella 180, 191, 252
Noctule 87
Nocturnal butterfly 124
Norfolk Island pine 135
Northern crested newt 13, 22, 66, 73, 92, 97
Northern cricket frog 38
Northern gannet 15, 88
Northern lapwing 16, 35, 39 f., 67, 99–101
Norway lemming 93
Norway maple 134, 138, 151, 163
Norwegian fjord horse 19
Nutria 48

O

Oak fern 133, 168
Oak processionary 95
Oak 134, 138 f., 141, 143, 150, 152, 155, 162,
    172 f., 185 f., 251
Oats 135, 140 f., 143, 145, 147–149, 154 f.,
    161, 165, 172 f., 175, 177 f., 180,
    183–185, 188, 234, 254
Ocean sunfish 12
Octopus 6, 54, 64, 123
Oil palm 156 f., 176
Old World monkey 43
Old World porcupine 45
Oleander 135 f.
Olive tree 134 f., 140, 156 f., 186

Olive 176 f.
Onager 19
One-flowered wintergreen or pyrola 171
Onion 146 f., 155, 173, 180–182, 188, 230,
    233, 251, 255
Opalina 94
Opium poppy 137, 146, 149, 161
Opossum 45, 76, 93, 105, 107, 114
Orach(e) 166, 170
Orange oak bolete 144, 169
Orange tip 95
Orange 182, 230, 233
Orangutan 8, 17, 47, 92 f., 107, 110 f., 114,
    239
Orca 17, 36, 78, 86
Orchard grass 143, 161, 165, 176
Orchid 160 f., 171
Ophrys genus 164
Orchidales 133
Oregano 153
Orinoco crocodile 14
Orpine (family) 159
Orthomyxoviruses 195
Oscillatoria 182
Ostrich 7, 16, 36, 55, 67, 69, 73 f., 79, 84, 86,
    93, 99 f., 102, 114
Owl 7, 34, 57, 119
Oxeye daisy 146, 166
Oxlip 167
Oyster mushroom 144, 169
Oyster 6, 31, 232, 234
Oystercatcher 7, 102

P

Painted comber (fish) 86
Painted quail 16
Palaeosiren 20
Palisander 138
Palm 152
Palmate newt 13
Palmleaf cinquefoil 182
Pansy 146, 165, 170, 174
Panther 69
Paper wasp 11, 51
Papovaviruses 195
Paramecium 54, 58, 87, 94, 252
Paramyxoviruses 195
Parasol 144, 169

Parmelia 187
Parnoviruses 199
Parrot 115, 126
Parsley 146 f., 155, 167, 176, 233
Parvoviruses 195
Pasque flower 146, 166
Passion flower 137
Pasteurella 197
Pea (family) 160
Pea 135, 140, 145, 147–149, 155, 166,
     171–173, 175, 177 f., 187, 253, 255
Peach 141, 146, 163, 181
Peacock 95, 127
Peanut 138, 156 f., 171, 175–177, 230
Pear (tree) 145, 162, 171
Pear 133, 137, 139, 179
Pearlwort (sagina) 161
Peat moss 142, 187
Pecan 176
Pectoral sandpiper 39
Peking duck 258
Pelargonium 141, 161
Pelican 7, 34, 123
Pelomyxia 94
Penguin 7, 88, 124, 258
Penicillium 197–199
Pennycress 166, 170
Pepper (vegetable) 230
Peppermint 146, 153, 167
Peppery milkcap 144, 169
Perch 74
Percheron 19
Peregrine falcon 34 f., 37, 39, 101 f., 121
Periwinkle 153
Pestilence wort 153, 167
Pheasant 7, 34, 50, 66, 76, 84, 99–102, 113,
     115
Phobosuchus 20
Picornaviruses 195
Pig 23 f., 30, 45–48, 50–54, 56, 61, 63, 69–76,
     78 f., 81, 90, 92, 105–107, 109–111,
     114 f., 118 f., 126, 250, 258
Pike 6, 12, 47, 66, 69, 71, 87, 90, 92, 96, 256
Pill millipede 36
Pilobolus fungus 174, 197
Pincushion flower 167
Pine chafer 122
Pine hawk moth 95

Pine marten 45
Pine moth 36, 95
Pine 133–135, 137 f., 142 f., 146, 150 f.,
     154 f., 180, 184 f., 255
Pineapple 182, 184, 187
Pinesap 166
Pink (family) 160
Pinniped 81, 120
Piperales 132
Plagiothecium 183
Plaice 6, 64, 81, 96, 121
Planarian 6, 36
Plantain (family) 160
Platanus 134, 137, 152, 163
Platypus 88, 90, 93
Plum tree 138
Plum 146, 163, 171, 230
Plumose anemone 117
Poales 133
Poinsettia 162
Pointer 18
Poison hemlock 167
Polar bear 8, 17, 36, 67, 88, 93, 105, 107
Polecat 45, 107, 115
Poliomyelitis virus 199
Polygonales 132
Pomeranian 18
Pond mussel 6, 30, 32, 54, 68, 80 f.
Pond-lily 167
Pondweed 164, 181
Poodle 18
Popcorn 172
Poplar 140, 146, 181
Poppy (family) 160
Poppy (genus) 137, 148, 171, 173, 176 f.
Porcini 145, 169, 230
Porpoise 47, 120
Potato 136, 140, 143, 146, 155–157, 166,
     173, 178, 180–182, 185, 188, 230, 233
Pox viruses 195
Praying mantis 6, 10, 81 f., 91
Prickly pear cactus 185
Primates 62, 121
Primitive insects 51, 68
Primrose (family) 160
Primrose 141, 146
Primulales 132
Prionosuchus 20

Privet hawk-moth 33, 68, 95
Privet 135, 146, 163, 186
Proconsul 239
Prokaryotes 132
Prosimian 45, 115
Proteus 196 f.
Protozoa(n) 5, 8, 94
Przewalski horse 19
Pseudomonas 196, 198 f.
Pteranodon 20
Pterosaur 20
Pterosphenus 20
Puccinia 199
Puff adder 27
Pug 18
Puma 108
Pumpkin 137, 142 f., 146–149, 154 f., 173, 179–181
Purple foxglove 161, 166, 173
Purple loosestrife 165
Purple moor grass 141, 165
Pussytoes 166
Pygmy mouse 17, 41
Pygmy salamander 13
Python 47, 54, 66, 90, 98

Q
Quackgrass 153, 165, 173
Quagga 19
Quaking grass 165
Quetzalcoatlus 20
Quince 163
Quoll 90

R
Rabbit 8, 17, 23 f., 31, 45–48, 52–54, 56 f., 60–63, 69, 71, 74–76, 78–81, 85 f., 90, 92 f., 104, 107, 109–111, 115, 118, 123, 126, 250, 253, 255, 258
Raccoon 45
Racehorse 36
Radiolaria 87
Radish 147, 149, 161, 230
Rafflesia arnoldii 137
Ragged robin 136, 166
Rainbow trout 22, 58, 64, 70, 96
Raisin 230
Ramaria aurea 144, 169

Ramie 252
Rampion 167
Ramsons 164
Rannoch rush 164
Ranunculales 132
Rape 135, 146, 148, 171–173, 176 f., 179
Raspberry 163, 179
Rat 8, 24, 30–32, 36, 45–48, 51–54, 56, 60, 62, 63, 68, 71–76, 79 f., 84–87, 90–93, 105–110, 114, 116, 118, 121, 123, 125, 253–256
Rattlesnake 7, 27, 69, 97, 258
Ray 80, 117
Red algae 132, 181, 183, 187
Red beet 147, 230
Red clover 135, 146, 148, 162, 167, 173, 179, 185
Red cracking bolete 169
Red deer (Wapiti) 17, 21, 36 f., 107, 112, 121
Red kite 103
Red underwing 95
Red-backed shrike 38, 103
Red-throated loon or diver 7
Redwood (sequoia) 134 f., 137, 151 f., 162
Reed bunting 103
Reed canary grass 162, 165
Reindeer lichen 183, 187
Reindeer 8, 17, 56, 107, 110
Reoviruses 195
Reptiles 5, 7, 14, 20, 26, 66, 76, 87, 92, 97 f.
Restharrow 163
Reticulated python 15
Retroviruses 195
Rhabdoviruses 195
Rhamphosuchus 20
Rhesus macaque 57, 62, 74, 84, 90, 92, 105–107, 111, 121, 124, 258
Rhinoceros 8, 21, 36, 44, 68, 93, 107, 110 f.
Rhizobium 196
Rhododendron 155
Rhubarb 182, 230
Ribbon worms 9
Riccia 187
Rice 143, 156 f., 175 f., 178
Rickettsia 196
Ringed seal 36
Roach 96

Roadrunner 36, 38
Robin 7, 55, 99, 103, 121, 125
Robinia 137, 139, 151, 163, 171, 179, 186
Rock dove 7, 16, 24, 30, 34 f., 47, 50, 52 f.,
    55 f., 60, 69, 71–76, 78 f., 84, 86, 92,
    112, 119, 121–123, 125 f., 255
Rook 16, 39, 103
Rosales 133
Rose (family) 160
Rose 155
Rosidae 133
Rotifer 6, 9
Rottweiler 18
Roundworm 5, 6, 9, 54, 77, 90 f., 249
Roxburghia 190
Royal albatross 101
Royal fern 168
Rubber tree 180
Ruby-throated hummingbird 35
Rudbeckia 162
Rudd 96
Rufous hummingbird 55
Rugosa rose 162
Runner bean (scarlet runner) 141, 172, 190
Ruscus 186, 190
Rush (family) 159
Russell's viper 27
Rye 135, 143, 147–149, 152, 154 f., 165, 171,
    173, 175, 183, 185, 188, 255
Rye-grass 162, 176

S

Sable antelope 21
Saccharomyces 197–199
Sacred lotus 173
Saffron crocus 161
Saffron milk cap 169
Sainfoin 166
Salamander 71 f., 80, 93, 122
Salmon 6, 12, 36, 66, 71 f., 96 f.
Salmonella 196–198
Saltwater or estuarine crocodile 14
Sand hopper 64
Sand lizard 14, 58 f., 69, 73, 92, 98, 121
Sandbox tree 174
Sandspurry 158
Sandtiger shark 97
Saprolegnia 199

Sarcina 199, 257
Sarcodon imbricatus 144, 169
Sargassum 182, 187
Sauerkraut 230
Sawfish 12
Saxifrage (family) 160
Saxifrage 153, 161
Scabious 165
Scaly wood mushroom 169
Scarlet pimpernel 158
Scentless mayweed 170
Scorpion 9, 26, 77, 82, 91
Scotch broom 162, 186
Scotch pine 163
Scrophulariales 133
Scyphozoa 8
Sea anemone 6, 8, 54
Sea club-rush 164
Sea cucumber 6, 9, 80
Sea holly 141, 167
Sea horse 6
Sea lamprey 12, 72, 96
Sea lettuce 181 f., 187, 252
Sea lion 8, 47, 61
Sea milkwort 153
Sea mouse 80
Sea or Shark catfish 96
Sea otter 88
Sea snakes 36
Sea urchin 6, 9, 54, 90 f., 253, 256 f.
Sedge (family) 160
Sedge warbler 40, 103
Sedge 146, 153, 164
Sedum 140, 161, 166, 185
Segmented worms (annelids) 5, 9, 87
Self-heal 165
Sepia 117
Serratia 196 f.
Sesame 157, 176
Seychelles giant tortoise 14
Shaggy ink cap 144, 169
Shark 20, 22, 36, 62, 71 f., 78, 80, 114, 120,
    124
Sharp-nosed ridged frog 38
Sheathed woodtuft 145, 169
Sheep 8, 23, 45–49, 52–54, 56, 61–63, 69,
    71 f., 74–76, 79, 81, 86, 90, 92 f., 105 f.,
    109–111, 114 f., 118, 250, 256–258

Shepherd's purse 142, 146, 166, 170 f., 255
Shetland pony 19
Shore crab 36, 54, 63 f., 77, 80 f., 122
Short-tailed shearwater 39
Shrew 8, 23, 30, 41, 45, 54, 56, 60, 68 f., 86, 93
Shrimp 64, 68, 120, 123, 232
Siberian brown bear 108
Siberian dogwood 163
Sickener 169
Silkworm 33, 80, 91, 94, 258
Silver fir 137, 151, 163, 183
    see also Fir
Silver ragwort 141, 153, 166
Silverfish 91
Silver-washed Fritillary 95
Siren („sea cow„) 46, 88
Sisal 156 f.
Skylark 99–102, 120
Slider turtle 71, 76
Slime mold 132, 249
Sloth 17, 31 f., 37, 85
Slowworm 7, 14, 22, 58 f., 69, 72 f., 80, 92, 97 f.
Small scorpion vetch 167
Small tortoiseshell 95
Small white (butterfly) 33, 37, 68, 91, 95, 250
Smooth meadow-grass 165
Smooth newt 7, 13, 59
Smooth snake 15, 97
Snails 5, 9, 87
Snake 14, 20, 31, 47, 125 f., 256
Snakefly 10
Snake's head fritillary 161, 164
Snapdragon 179
Snapping turtle 258
Snipe 39 f., 102
Snow flea 58
Snowdrop 141, 146, 164, 188
Soapwort 153, 167
Sod 181
Soft-shell clam 51
Soldier beetle 33
Sole 73
Solid-stem burnet saxifrage 165
Solomon's seal 164
Somali wild ass 19

Song thrush 39–41, 103
Songbirds 22, 93
Sorrel 146, 155, 167
Sour cherry 163, 179
Soy bean 156, 175–177
Sparrow 23, 35, 52, 54 f., 69, 71, 76, 79, 86, 103, 105, 114, 120, 125, 254, 256
Sparrowhawk 16, 37 f., 56, 103
Spearmint 153
Sperm whale 17, 46 f., 68, 78, 88
Spermatophytes 132
Spider 6, 9, 26, 37, 82, 91
Spider monkey 114
Spiderworts (*Tradescantie*) 141, 142, 180, 255
Spikemoss 146, 168
Spinach 146 f., 149, 162, 167, 177, 188, 230, 233
Spine-tailed swift 37
Spiny mouse 108
Spirochetes 196
Spirogyra 181
Sponge 5 f., 8, 54, 91
Spotted flycatcher 102
Spring barley 149, 172
Spring rye 149
Spring snowflake 164
Spring wheat 149
Spruce 134 f., 137, 145, 150–152, 154 f., 163, 171, 183–185, 188
Spurge (family) 160
Square-lipped rhinoceros 17, 21
Squid 64, 123
Squirrel 8, 17, 45, 48, 52, 59 f., 67, 106, 115
Squirting cucumber 174
St. Bernard 18
St. John's wort 166, 170
Stag beetle 68
Staphylococcus 196, 198
Star of Bethlehem (Gagea villosa) 164
Starfish 6, 9, 37, 54, 77, 80, 91
Stegosaurus 20
Steinheim man 239
Stellaria 142, 167, 185
Stentor protozoan 91
Stentor 94
Sticklebacks (gasterosteidae) 64, 96, 254
Stingray 52

Stinkhorn 145, 169
Stone bramble 153
Stonecrop 166
Stoneflies 10
Stony coral 6
Stork 7, 16, 34 f., 37, 39 f., 56, 67, 69, 84, 86,
    99 f., 102 f.
Strawberry 138, 141 f., 145, 153, 155, 162,
    166, 230, 233
Strawflower 167
Streptococcus 196–199
Sturgeon 6, 96 f.
Stylonychia 94
Subterranean termites 11
Suede bolete 169
Suffolk 19
Sugar beet 157, 181, 185
Sugar kelp 182
Sugarcane 152, 156 f.
Sulphur tuft 145, 169
Summer fruit tortrix moth 95
Sundew (family) 160
Sundew 146, 167
Sunflower 136, 138, 141 f., 144, 146–149,
    152, 154 f., 167, 176 f., 179, 181, 184 f.,
    188, 255
Swallow 7, 37, 67
Swallowtail butterfly 33, 95
Swallow-wort 135, 167
Swan orchid 171
Swan 7, 22, 37, 67, 84
    see also Mute swan
Sweet chestnut 134, 137, 139, 152, 162, 173
Sweet clover 167, 179
Sweet corn 138, 172
Sweet gale 163
Sweet pea 178
Sweet potato 156 f., 178
Sweet vernal grass 165
Sweet william 147 f.
Swiss pine 134 f., 137, 163, 183
Swordfish 12, 36
Swordtail 92
Sycamore maple 133 f., 137, 150, 162
Syrian hamster (or Golden hamster) 7, 17,
    52, 61, 69, 71, 73 f., 76, 78, 84 f., 92 f.,
    104, 106–109, 114, 120, 254

T
T2 phage 257
Tabanus bovinus 37
Tall oatgrass 165
Tapeworm 6, 8
Tapir 45
Tarantula 6, 26
Tarsier 92
Tawny owl 16, 56, 103, 118 f., 125
Tea 156 f.
Teasel (family) 159
Tench 12, 250
Termite 6, 10 f.
Terrestrial nematodes 6, 54, 250
Thiospirillum 251
Thornback ray 76
Three-toed sloth 22
Thrush 57
Thyme 167
Tiger beetle 37, 58
Tiger shark 12, 66
Tiger snake 27
Tiger 8, 17, 38, 41, 45, 47, 63, 93, 108, 111,
    120
Tilia (linden, lime) 134, 139, 143, 152, 155,
    177, 179, 186
Timothy 161, 165, 188
Tinder polypore 145
Titanichthys 20
Toad 31 f., 52, 93
Tobacco mosaic virus 251
Tobacco 142, 144, 146 f., 154–157, 162, 170,
    173, 181, 188, 255
Togaviruses 195
Tomato 138, 141, 146 f., 167, 181 f., 188,
    230, 233
Toothed wrack 187
Tormentil 165
Tortoise 7, 58, 97
Touch-me-not 167, 185
Tree frog 7, 13, 59, 66, 92, 97 f., 124, 256
Tree heath 186
Tree pipit 40, 102
Trematode 8
Trichina 6
Trout 6, 12, 36, 54, 66, 71, 93, 96, 119, 123,
    231, 234, 256 f.
    see also Rainbow trout

True bugs 5, 10
True grass (family) 160, 165
Truffle 169
Tuatara 7, 14, 93
Tufted hairgrass 165
Tufted vetch 167
Tulip 142, 188
Tuna 12, 36, 66, 114, 231, 234, 258
Tunic flower 158
Tunicate 68
Turbellaria 8
Turbot 96 f.
Turkey 34, 69, 99, 258
Turnips 182
Turtle dove 39
Turtle 14, 20, 22, 31, 46 f., 62, 70, 72, 75,
    78 f., 256
Twayblade 164
Two-toed sloth 22
Tyrannosaurus 20

U
Umbellales 133
Urticales 132

V
Vaccinium 133, 146, 163
Valerian 145, 165
Vallisneria 180
Valonia 191
Veiled mushroom 150
Velvet bolete 169
Velvet grass 165
Venus flytrap 146, 191
Veronica 161, 166
Vertebrates 5 f., 22 f., 25, 28, 46, 52–54,
    58–60, 62, 66, 69, 72–74, 76–79, 84 f.,
    88, 92, 114, 120, 123, 126

Vertebrates, extinct 20
Vetchling 167
Vibrio 196–198, 257
Victoria regia 136 f.
Violales 132
Violet (family) 160
Violet 142, 144, 146 f., 153, 155, 162, 174
Viper's bugloss 143, 167
Viroid 195

Virus 195, 198, 251, 257
Vorticella 94

W
Walking sticks 10, 51, 80, 94
Wall-rue 168
Walnut 133, 137, 140, 142, 155, 163, 171,
    173, 176, 186, 230, 234
Walrus 17, 45, 73, 84, 88, 108
Wart-biter bush cricket 127
Warthog 21
Wasp 11, 33, 58
Water arum 164
Water avens 165
Water buffalo 17, 21
Water chestnut 167
Water flea 6, 77, 91, 122, 250
Water hemlock 167
Water lily family 160
Water lily 137, 141 f., 144, 158, 167, 188
Water parsnip 166
Water plaintain 164
Water scavenger beetle 34
Water vole 17, 58
Watercress 166
Watermelon 138, 230
Watermilfoil 181
Waterwheel plant 191
Wayfaring tree 163
Weasel 68
Weddell seal 88
Wels catfish 12, 96
Welwitschia 135
Western diamondback rattlesnake 15
Western green lizard 14, 37, 66, 69
Whale shark 6, 12
Whale 8, 56, 62, 69, 90
Wheat 135 f., 141–144, 146–148, 154–157,
    165, 172 f., 175–177, 181, 183–185,
    188, 255, 257 f.
Wheatgrass 161
White clover 173, 179
White or elfin saddles 169
White poplar 137, 163
White wagtail 40, 99, 102
Whitebeam 163
Whitefishes 96
White-lipped banded snail 91

White-tailed eagle 16
Whitethroat 40, 102
Whitlow grass 136, 166
Wild ass 37
Wild barley 165
Wild boar 8, 17, 93, 105, 108, 111 f., 115
Wild cat 45, 105
Wild cherry 163, 179
Wild chervil 144, 167
Wild geese 112
Wild horse 37
Wild mustard 161, 170, 185
Wild oat 165
Wild radish 145, 166, 170
Wild rosemary 163
Willow (family) 160
Willow warbler 40, 102
Willow 140 f., 144, 146, 163, 186
Willow-herb 161
Wine grape 136, 146, 155, 156, 163, 181,
    182, 184
Winged broom 163
Winter aconite 161
Winter barley 149, 162
Winter cress 165
Winter moth 58, 95
Winter rice 162
Winter rye 149, 172
Winter wheat 149, 162, 172
Winter wren 7, 23, 56, 67, 99, 101–103
Wolf 8, 17 f., 37, 45–47, 52, 84, 86, 93, 108,
    111, 114
Wolverine (Glutton) 17
Wood anemone 136, 140 f., 166
Wood ant 11, 122
Wood betony 168
Wood blewit 169
Wood grouse or Capercaillie 16, 99, 112
Wood manna-grass 165
Wood mouse 8, 38, 60, 68, 108
Wood pigeon 16, 38, 67, 99 f., 102 f.
Wood sanicle 167

Wood small reed 165
Wood sorrel (family) 160
Wood warbler 40, 103
Woodland horsetail 168
Woodlark 40, 102
Woodlouse 54, 68, 91, 122
Woodruff 153, 167
Woodrush, Great 164
Woodsorrel 141 f., 167, 183
Wormseed mustard (Wormseed wall-
    flower) 165
Wryneck 40, 103

Y
Yacare caiman 124
Yak 110
Yams 157, 178
Yeast 251, 256–258
Yellow archangel 166
Yellow fever virus 198
Yellow flag 164
Yellow gentian 145, 166
Yellow goat's beard 158
Yellow mealworm beetle 91, 94
Yellow morel
Yellow or common toadflax 166
Yellow Star-of-Bethlehem 140
Yellow woundwort 185
Yellow-bellied toad 13, 98
Yellowcress 167
Yellowhammer 99, 102, 120
Yellow-rattle 166
Yew 134 f., 138, 142, 145, 151 f., 162, 183,
    186

Z
Zander 12, 97
Zebra 8, 37, 115
Zebu 111